IMAGES
of America

COLFAX

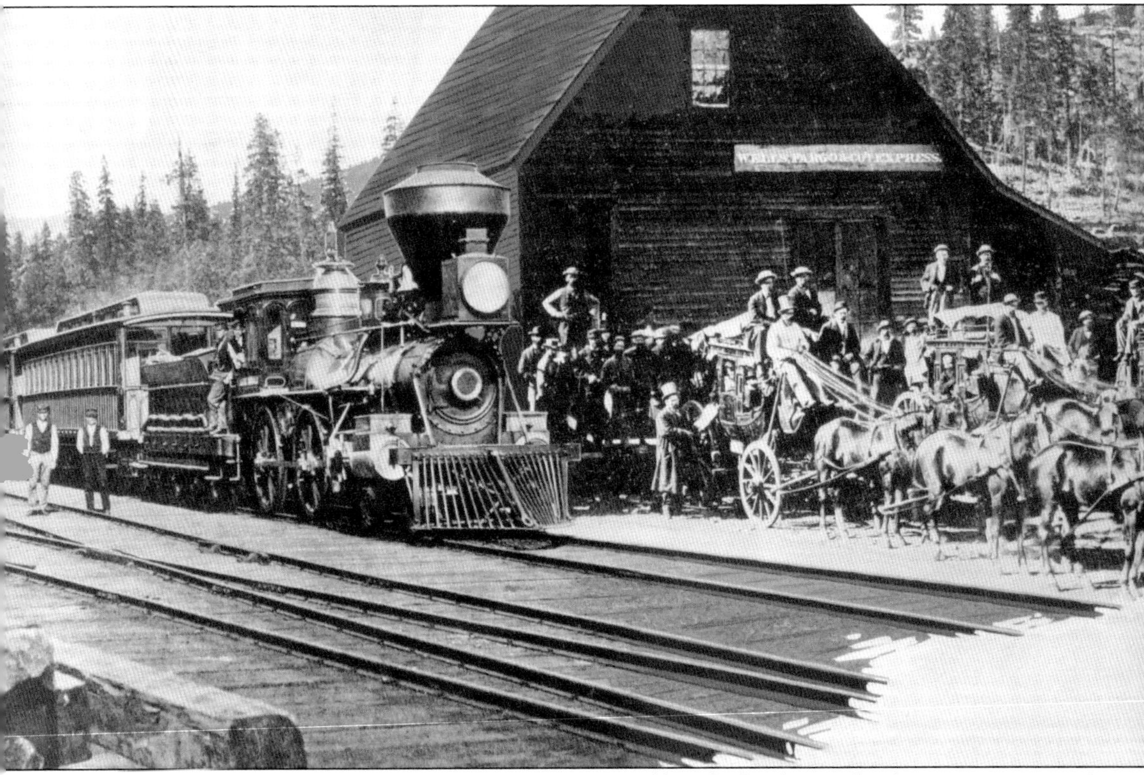

This marvelous 1884 photograph of the train, stopped beside the depot, clearly portrays a typical scene in the northern foothill towns (Cisco Grove). The Colfax train station also housed the Western Union Telegraph, and the passenger depot, plus a restaurant (see page 32). It served as a transportation hub for the northern foothills. Passengers disembarked from the train, and those continuing on climbed into the awaiting stagecoach towards their final destinations in the smaller communities of the foothills. The Colfax station burned in 1905 but was replaced. (Courtesy of J.L. Johnson Photography.)

ON THE COVER: In 1900, Colfax still had some wooden structures surrounding these first brick buildings on Main Street. The dirt streets and wooden sidewalks are typical of Old West towns and reminiscent of those seen in early Western movies. (Courtesy of J.L. Johnson Photography.)

IMAGES
of America

COLFAX

Jan Westmore

ARCADIA
PUBLISHING

ISBN 978-0-7385-9611-2

Published by Arcadia Publishing
Charleston, South Carolina

Printed in the United States of America

Library of Congress Control Number: 2012941474

For all general information, please contact Arcadia Publishing:
Telephone 843-853-2070
Fax 843-853-0044
E-mail sales@arcadiapublishing.com
For customer service and orders:
Toll-Free 1-888-313-2665

Visit us on the Internet at www.arcadiapublishing.com

This book is dedicated to all the hardworking volunteers at the archives and the museums that give selflessly of their time in the love of history and to preserve its richness for the generations to follow.

CONTENTS

Acknowledgments

Special appreciation goes to Arthur Sommers and Jim Johnson, who opened up their private photograph collections. With their generosity, the visual dimension of Colfax's history was given life. The Library of Congress proved a valuable tool, as online images are offered free to all.

Many individuals supported my project in various ways. Debbie Polson at the Placer County Museums–Archive and Research Center instructed me on how to access the numerous resources in the archives. Helen Wayland and Donna Williams at the Colfax Heritage Museums facilitated my search for historical photographs. The staff at the Golden Drift Museum in Dutch Flat offered resources covering neighboring towns.

Helen Wayland, Roger Staab, Nancy Hagman, April Moore, Sally Knutson, and Claudine Chalmers provided historical insight and local lore. The Searles Library in Nevada City and the Grass Valley Museum provided access to important historical references. Railroad museums in Nevada County and Sacramento steered me through their exhibits, providing important details about the transcontinental and narrow-gauge railroads and how freight and passenger transportation contributed to the development and success of the area in the 19th and 20th centuries. I would like to express my thanks to the Sierra Writers and the associated nonfiction Critique Group for their ongoing support.

The images in this volume appear courtesy of Claudine Chalmers (CC), Colfax Heritage Museum (CHM), Doris Foley Library (DFL), Grass Valley Museum (GVM), J.L. Johnson Photography (JLJ), Sally Knutson (SK), the Library of Congress (LOC), Placer County Museums (PCM), and Arthur Sommers (AS).

INTRODUCTION

Once word of the discovery of gold in California echoed around the globe, pioneers rushed en masse to the Sierra Nevada foothills seeking their fortune and hoping for a bright new future. They imagined getting rich quick by merely scooping up nuggets by the handfuls out of mountain streams. The scramble to realize those dreams caused tiny settlements to sprout up along every riverbed and ravine and up the mountainsides.

Native Americans who lived in the area were all but wiped out by the pioneer invasion and the towns that usurped their territory. Newcomers rapidly seized land and consumed resources, displaying greed and an attitude of manifest destiny. Indians died from foreign diseases, starvation, and outright decimation. Others were displaced. Most settlers held no respect for the Indians and their different way of life. Instead, the native presence was seen as an impediment and a threat.

Few miners found enough gold picking or panning to sustain them, let alone get rich. The work was physically demanding and often yielded little. Those discouraged miners with an entrepreneurial spirit abandoned the pursuit of a chance fortune. Instead, they began "mining the miners," says Colfax historian Nancy Hagman. Merchants sold the miners the tools needed to chase their dreams—simple things like picks and shovels, pans and coveralls, and of course, good food. The price of food soared after the discovery of gold in California. In 1849, a slice of bread might cost $1—and a buttered slice might cost $2. Potatoes and onions were $1.50 per pound. The women cooking stews and baking biscuits fared very well, as they could set their own prices, which miners gladly paid, savoring every "home-cooked" bite.

Illinoistown, a half-mile from the future Colfax, became the major supply distribution center from ports in San Francisco and Sacramento. Many items also came overland by mule teams and stagecoaches. Wares were then distributed to merchants in remote towns and to miners at their claims, scattered throughout the foothills. Hotels, forges, brothels, saloons, and general stores offered both the goods and entertainment opportunities the miners coveted. Farmers also found the area's climate conducive to planting orchards along with a willing market for produce.

Some families moved to the area during the Gold Rush, but men initially dominated the populace. Once the frantic explosion of individual gold-seekers waned, the towns that would eventually survive began to stabilize as additional women and children arrived. More churches, schools, and meeting halls appeared. Townspeople constructed buildings from timber, which was plentiful at the time. With so many wooden structures, fire was the cause of most 19th-century disasters. Volunteer firefighters, with their bucket brigades and limited resources, could not stop the infernos. However, the people would not be beaten, and towns that burned were rebuilt.

Eventually, locals tired of wooden buildings being destroyed and began building with bricks made by a company just up from the Methodist church on Church Street. Some wooden two-story Victorian homes with charming balconies remained for those who could afford them. Many of these jewels still stand today, sprinkled through the foothill towns.

While the gold ultimately provided income to town merchants and miners alike, the tedious extraction process underwent a rapid evolution in just the first few years after the initial gold discovery in Coloma. Mining shifted from individuals working the streams and rivers with gold pans to teams using wooden boxes on rockers that enhanced the sifting capacity. A miner in the foothills developed a system of hydraulic mining that further increased efficiency by directing a high-powered hose at the hills and embankments to wash away surface rocks. Placer County actually got its name from the miners' extensive use of this placer method in the area.

A significant problem arose from the use of hydraulic mining. Contaminated runoff dumped into the rivers then drained into the valley, flooding places like Marysville and polluting the rich soil that farmers needed for agriculture. In 1884, Judge Lorenzo Sawyer of the Ninth Federal Circuit Court decided, in *Woodruff v. North Bloomfield,* to ban hydraulic mining in California. By 1893, Congress was once again allowing the process, but only where the sediment could be detained. Hydraulic mining was never revitalized in the Sierra foothills again, since the infrastructure had been removed from most sites. Neither law focused on the massive destruction of the landscape caused by hydraulic mining.

Gold fever did not diminish with the ban on the hydraulic method, however. Mining companies redirected their pursuit of the precious metal. They began hard-rock mining, utilizing underground shafts and tunnels. Deeper and deeper they dug through the granite, hoping to locate layers of quartz laden with gold.

With all of the migration to the west of people and supplies, first seeking gold and then land, chatter escalated about connecting the eastern and western United States by transcontinental railroad. With the Civil War imminent, President Lincoln dispatched a cadre to investigate routes and possibilities. Speaker of the House Schuyler Colfax was part of that effort and took several trips across the country to determine the best route. Congress approved the railroad under the initial Pacific Railroad Act of 1862. The job was completed when the Golden Spike was driven on May 10, 1869, at Promontory Point, Utah.

Building the tracks over the Sierra Nevadas was a major feat that would never have been accomplished without thousands of hardworking Chinese immigrants. Under the direction of Charles Crocker, more than 10,000 Chinese arrived to take on the task. Central Pacific built a camp a half-mile up from Illinoistown, which became the staging area and dispatch center of labor and materials for laying tracks east across the country over the challenging Sierras. The local lumber industry thrived with the railroad construction demands.

Historians now believe that Gov. Leland Stanford actually named the town Colfax in a political move to gain project support for the railroad. The auctioning off of lots set the new town in motion. Merchants relocated from Illinoistown to avail themselves of the new market. Colfax took over as the key distribution point for supplies to mining towns and remote locations in the foothills.

Nevada City and Grass Valley, now thriving towns, relied on Colfax. They were "off track" and demanded a more efficient and direct supply method. Wagon roads were rough, rocky, and steep, and delivery methods could not handle the increasing demand for heavier shipments. Adding a new railroad to intersect the Central Pacific became essential. In May 1876, just two years after the route was surveyed, the Nevada County Narrow Gauge Railroad (NCNGRR) was completed, linking Nevada County and Placer County. The new railroad carried passengers and freight between the two interdependent areas.

The NCNGRR transported gold to Colfax, but it also facilitated the marketing of another valuable asset—agriculture. Both Placer and Nevada Counties were renowned for the quality of their fruits. Once the Central Pacific and the NCNGRR met, the opportunity to transport fruit from towns and farms to the western cities—and across the country—became a reality. Insulation improved on railcars, and in 1926, an icing platform was erected in Colfax. Similar structures were built at regular intervals to enable the transcontinental delivery of fresh fruit. Farmers brought their apples, pears, and peaches by wagon or via the NCNGRR to the fruit sheds in Colfax, where the Pacific Fruit Exchange packed the produce for transport by refrigerated railroad cars, known as "reefers."

Along with gold and high-quality produce, another area treasure was the climate. Residents loved to say that Colfax was "above the fog but below the snow"—true most of the time. Dr. Robert A. Peers, a Canadian physician, arrived from Ontario in 1899. Delighted at how conducive the climate was to good health, he opened a hospital in Colfax in 1907. He also established the Colfax School of Tuberculosis to treat the ever-growing number of afflicted patients. Dr. Peers also consulted at a sanatorium in nearby Weimar. Soon, there were several sanatoriums in the Colfax area, and people came from all over the country to be treated for the illness. Once the epidemic subsided with the advent of antibiotics in the 1940s, some of the facilities were converted into general hospitals.

Toward the end of the 1900s, Colfax became an attractive tourist destination. Today, visitors come by train and by automobile to see the historic town. Colfax serves as a stopover for travelers when visiting the foothills or en route to a longer vacation at one of California's many natural wonders, such as Donner Lake or Lake Tahoe. Colfax continues to focus on the synergy of the tourism trade and the railroad. Even locals in the neighboring communities occasionally enjoy hopping on the train for a jaunt to the big city or a short trip up to Reno on a winter day, when the snowfall overlooking the lakes is magical.

Amtrak now operates the railroad that rumbles into town each day carrying passengers to their destinations. The freight trains, some with "double-stack" cargo containers, chug through, transporting products and livestock from one state to another, to depots and ports all across the continent. The Colfax Heritage Museum, in the local train depot, lures folks to stay on and dive deeper into the lives and tales of the people of this charming old western town.

People were drawn to this foothills town by gold, stayed for the climate, prospered from the agriculture and timber, and then grew with the railroad.

One

Original Residents

Native Americans have lived in North America for thousands of years. The Nisenan branch of the Maidu considered the foothills of the Sierra Nevada home. Here, they fished, hunted, and gathered natural foods like acorns and pine nuts. In the hot summers, they relocated to traditional hunting grounds like the Gold Run and Dutch Flat areas.

After gold was discovered in the region, the onslaught of miners brought chaos and disease to the Nisenans. Never having been exposed to white diseases, Indians' immune systems caused many to fall victim to the newly introduced germs. Unlike the invading foreigners who seized without conscience, the native peoples honored the land as a source of life. They believed one does not own the earth but instead benefits from its many offerings. The fruits of the lands were to be prudently shared, not depleted, in order to preserve the gift for future generations.

To the miners and settlers, gold existed for the taking, at the expense of anyone or anything that got in the way. Mining claims supplanted the Nisenan-Maidu camps, and the people suffered through constant relocation and elimination. A local militia, the Blades of Colfax, sought Indian decimation. Finally, in 1917, the California Supreme Court declared Indians citizens, but it was not until 1924, under the Indian Citizen Act, that the federal government recognized all people born within the territorial United States as citizens.

A well-known local Nisenan-Maidu, Maria Kennett, is buried in the local Indian Cemetery in Colfax. She married a white man, William Kennett, after saving his life when he was trapped under a rock in the American River and nursing him back to health. Maria died in 1918 at about 89 years of age. Indians gathered from surrounding counties to honor her at "the cry" and the funeral.

About 1916, the federal government set aside a 40-acre parcel for an Indian Rancheria, where homeless Indians could live. The land was poor and had little access to healthy water. In the 1940s or 1950s, when none of the designated land was being occupied by Nisenans, the government released it for auction.

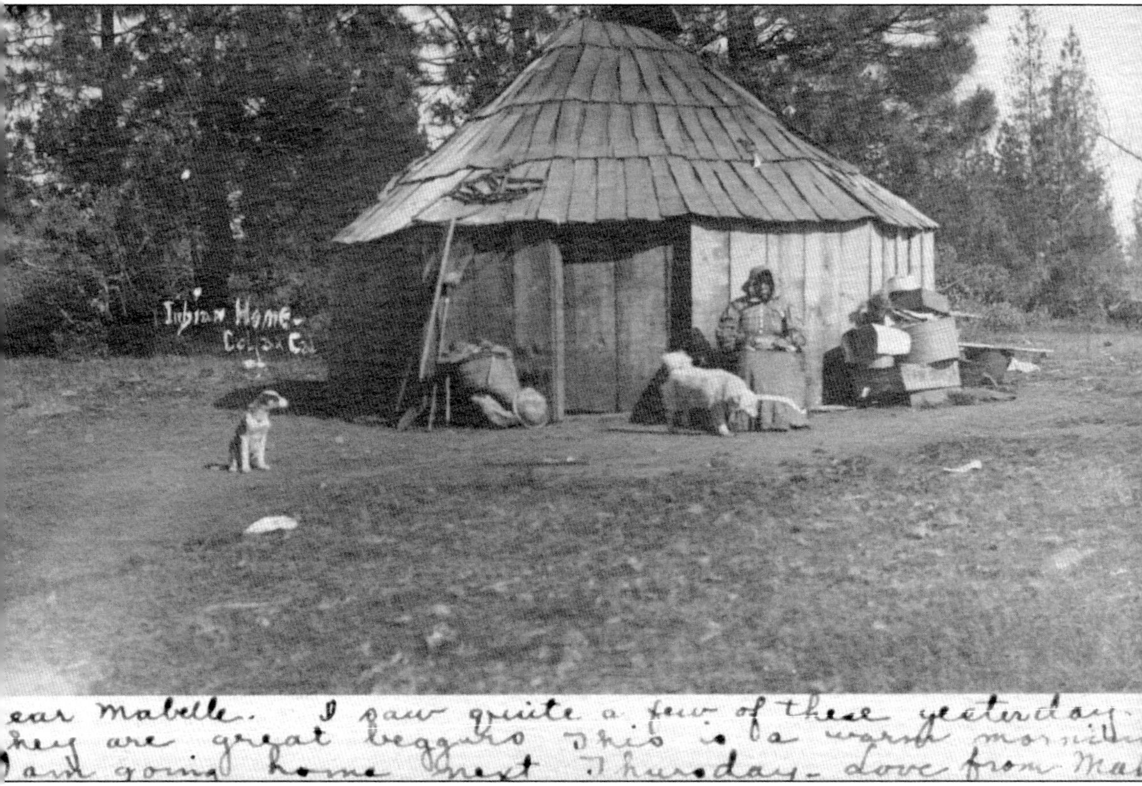

This postcard, mailed in 1906, shows a Native American woman seated outside of her home in Colfax. Some referred to it as a roundhouse, but it was a bark house, or a *hy*, constructed from the local cedars that were plentiful in the area. Indians would burn down these homes upon the resident's death. Customarily, when a person died, Native Americans hosted a potlatch, where they would give away or burn all of the deceased's possessions. Nisenans generally used a special building called a *kum*, or roundhouse, for tribal meetings and traditional dancing during ceremonies. The message on the card, from Mama to Maybelle, reflects a judgmental attitude of the times. (AS.)

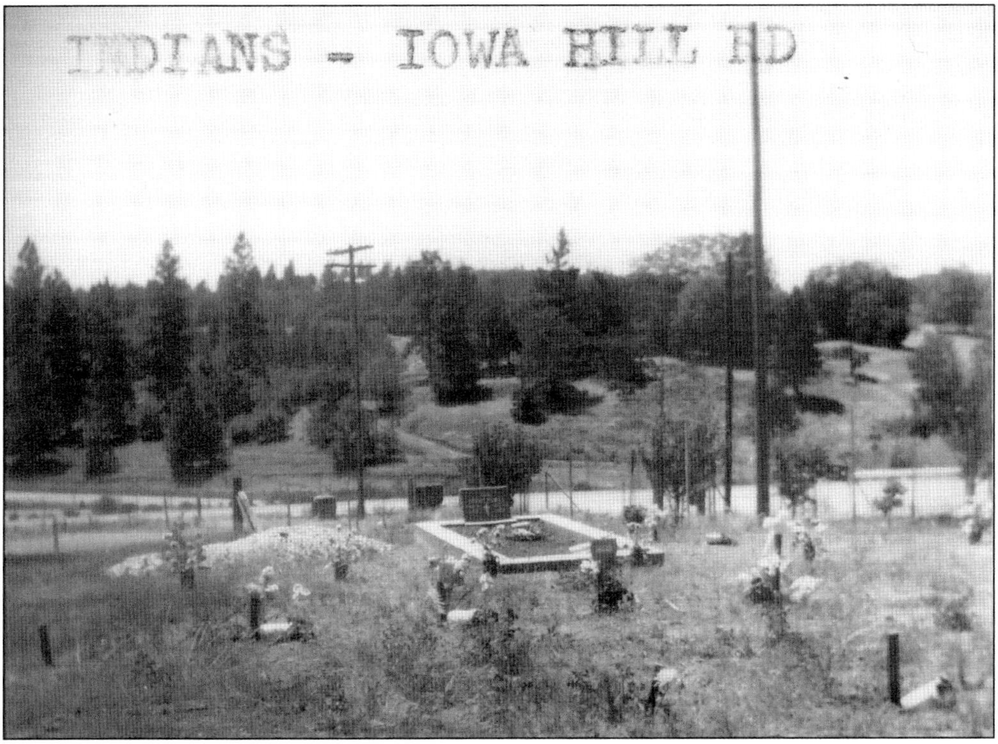

Indian burial grounds have always been sacred to the Native Americans as the final resting place of their people. Preservation of the grounds honors the ancestors and demonstrates support for the Indians' long-held traditions. The Indian Cemetery on Iowa Hill Road just off of South Canyon Way Road in Colfax is seen here in 1965. (PCM.)

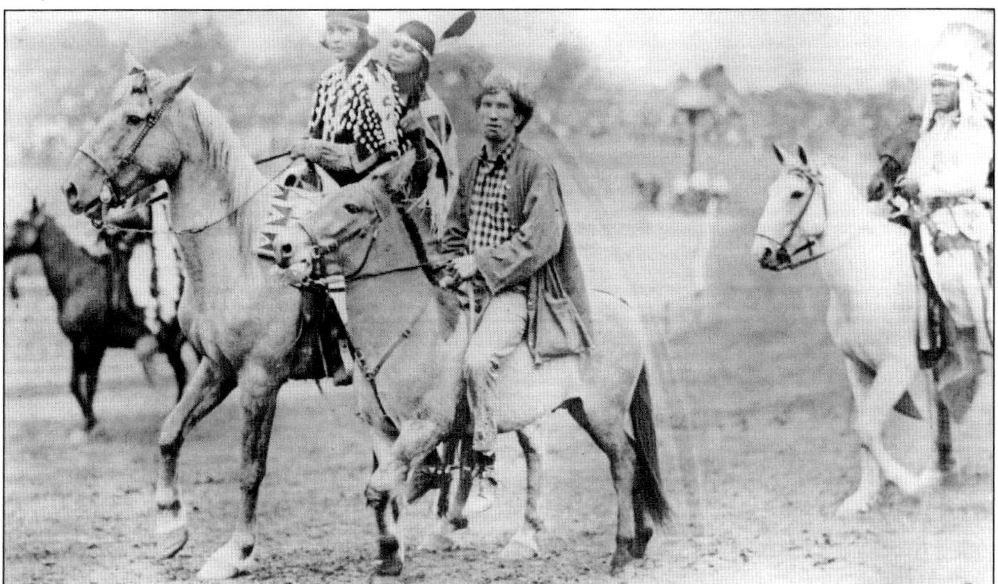

Native Americans ride horseback in a Colfax parade in 1930. The event was likely in celebration of the Fourth of July. Note the headdress on the man on the right edge of the image; although it is not traditional Nisenan, it is magnificent. (CHM.)

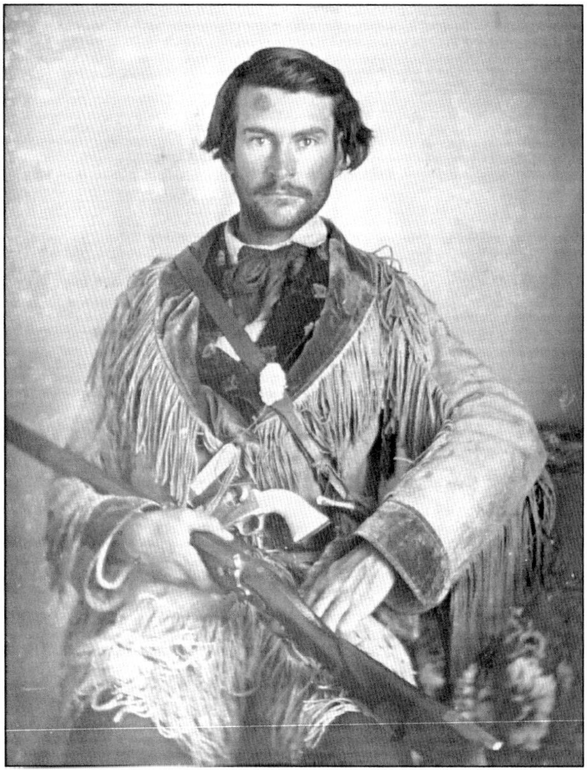

Simmon Peña Storms came to America after growing up in Venezuela. He left Boston, Massachusetts, and headed west with a group of 150 men, including David Bovyer, who became his employer upon his arrival in the Chicago Park area. With a knack for languages, Storms soon became a special courier there and a translator for Native Americans and foreigners. In 1852, Storms encouraged Chief Weimah to hold a *yomashee* at the ranch to educate the miners and merchants about the Native American traditions. (GVM.)

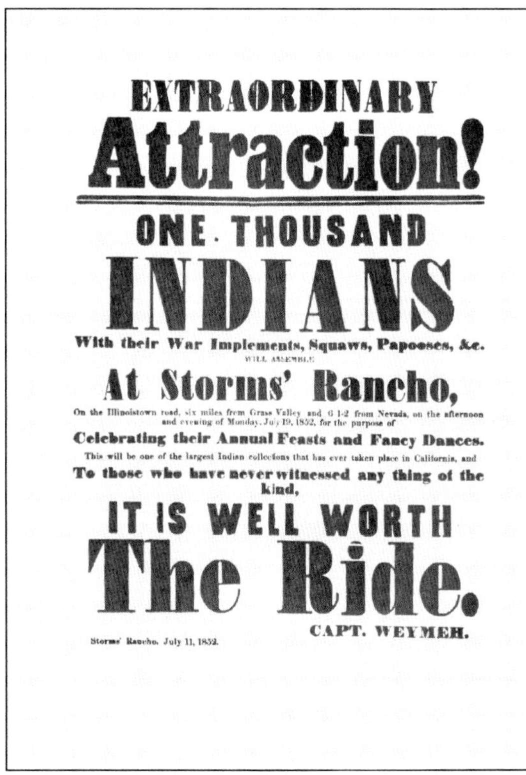

This poster dramatized the event Chief Weimah organized, hoping to attract people to the festivities at Storms's ranch on July 19, 1852. (GVM.)

Two

Mining Mania

At the onset, mining was done the hard way, using gold pans in the chilly rivers and streams. While few made their fortune panning for gold, many continued struggling at it up and down the Bear and American Rivers. Sluice boxes and then rockers helped to sort the gravel and isolate the gold.

Claims were jumped, depleted, and abandoned. When waterpower became available via flumes diverting it from the rivers, hydraulic mining became the most efficient way to extract the precious gold. Company operations soon dwarfed individual mining efforts. Men aimed the flow of the powerful nozzles called monitors at the hillsides, blasting them away with the high pressure. The resulting residual rock and debris was moved into huge sluices to isolate the gold.

When rivers became polluted and caused floods in the Sacramento Valley, the federal government ruled that hydraulic mining must cease, but nothing could stop the hunger for riches. Gold mining went underground. Even when cave-ins and the threat of silicosis threatened premature death, men keep going underground. They rode carts for miles through tunnels dug deep into the mountains hoping to uncover rich gold veins. Each wanted to locate the "mother lode," or main vein. Miners labored by hand using picks and later drills as well as explosives to break up the granite to reveal the quartz where the gold might be hidden. Ore-filled carts were raised to the surface and transported to stamp mills, where the ore was crushed and the gold was isolated with mercury.

The Rising Sun was the richest gold mine in Placer County. The hard rock mine produced 200,000 ounces of gold between 1866 and 1932. Tonnes (1,000 kilograms) of gold were taken from the foothills, but many believe that even more still lies waiting. The mines remained open during the Great Depression, keeping foothill economies afloat, but the federal government ordered the mines closed on October 8, 1942, under the War Production Board Limitation Order L-208 to preserve resources during World War II. The California mining companies never rebounded after the war, as inflation combined with the fixed price of gold undermined potential profit margins.

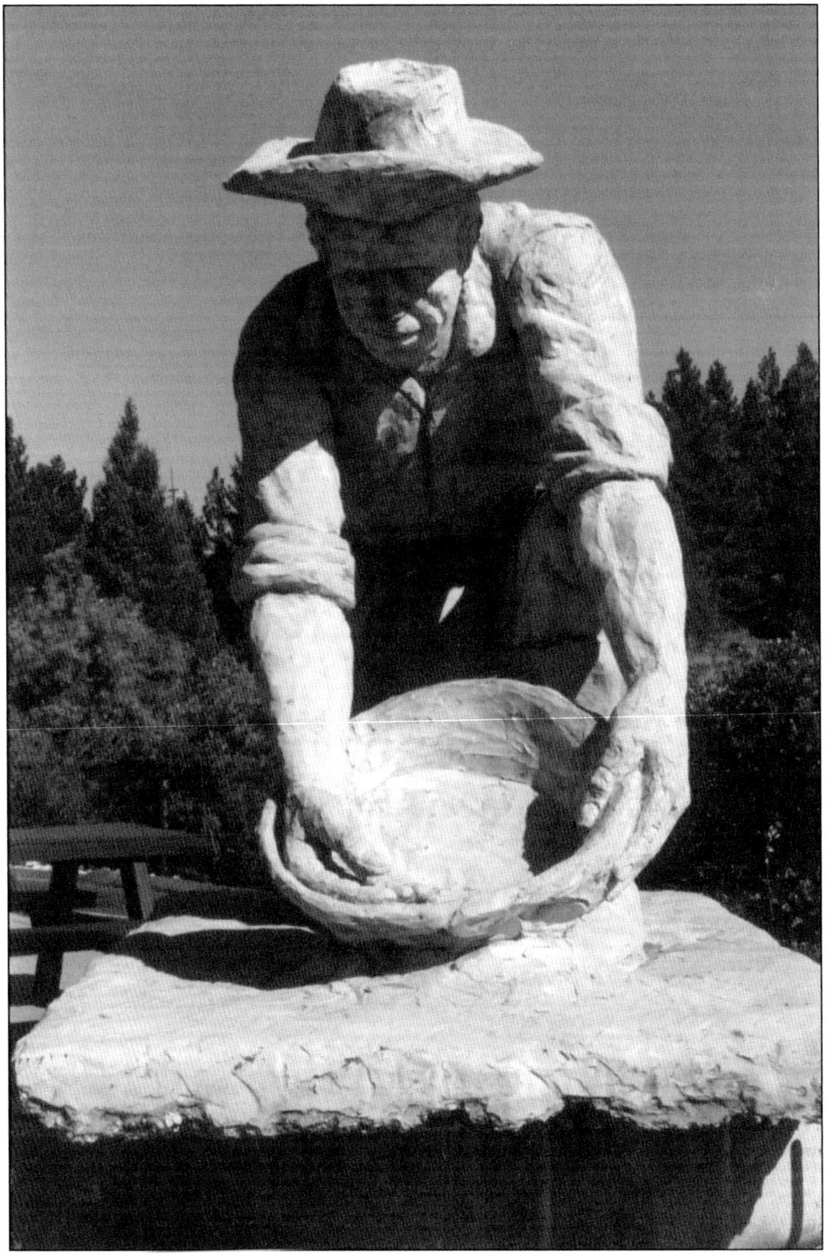

Many locals remember this concrete statue of a miner displayed in front of the Colfax Hotel. Artist Ken Fox created the representation of Claude Chana, a miner who discovered three large gold nuggets while panning near Auburn Ravine on May 16, 1848, not long after James Marshall's discovery in Coloma. Chana emigrated from France to America and then moved west. His wagon train encountered the Donner Party in Utah in 1846, but he chose a different route and avoided the horrendous Sierra storm. At the Johnson Ranch on Bear River, Chana met Theodor Sigard. They planted peaches and almonds at Sigard's ranch, but Chana caught gold fever. Just when he was about quit panning, he found the nuggets. Upon retiring with his riches, he returned to Bear River before buying a ranch in Wheatland. He went on to become one of the first California winemakers. The Colfax statue of Chana now resides in storage. (CC.)

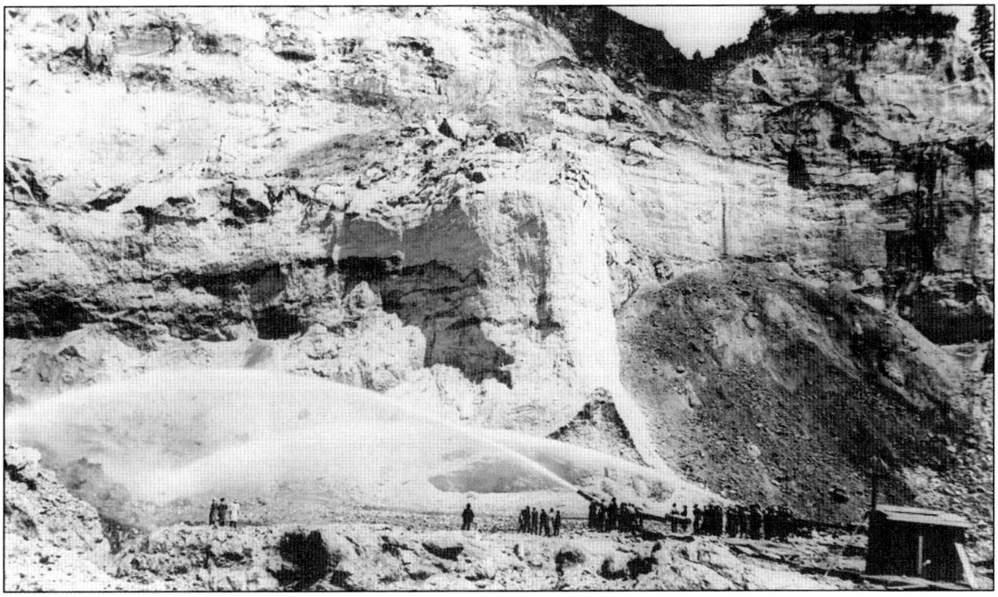

Men worked every day managing the monitors—hoses with giant iron nozzles—and directing the powerful jets toward the hillside bluffs. Water came from the Bear and Yuba Rivers and the North Fork of the American River. The force of the flow loosened the rock and sent it tumbling down to where the ore was separated and the gold extracted. It was a fast-acting remedy to pick-and-shovel mining, but the hydraulic method destroyed every speck of scenery in the way. (JLJ.)

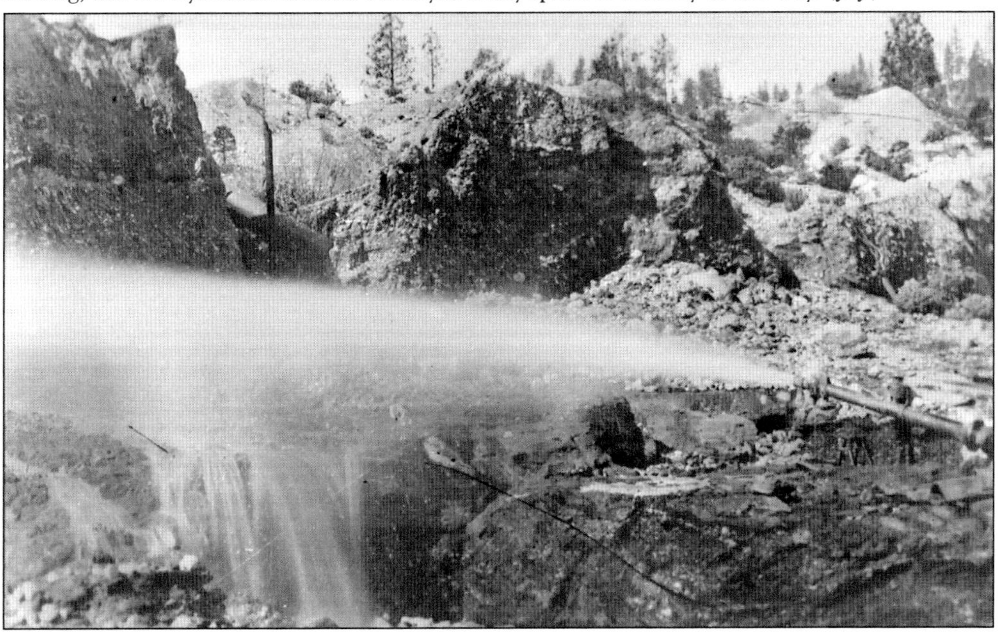

Ed Matteson first applied this high-powered method near Nevada City, California, in 1853. The tremendous blast of water pushed aside trees and foliage, crumbled hills, and created temporary waterfalls. The slurry, containing sand and rock mixed in with the gold, traveled downstream into sluice boxes four feet by four feet and about 1,000 feet long. There the mix encountered "sluice ruffles" and drainage tunnels, where gold combined with liquid mercury, extracting it from the rubble. (JLJ.)

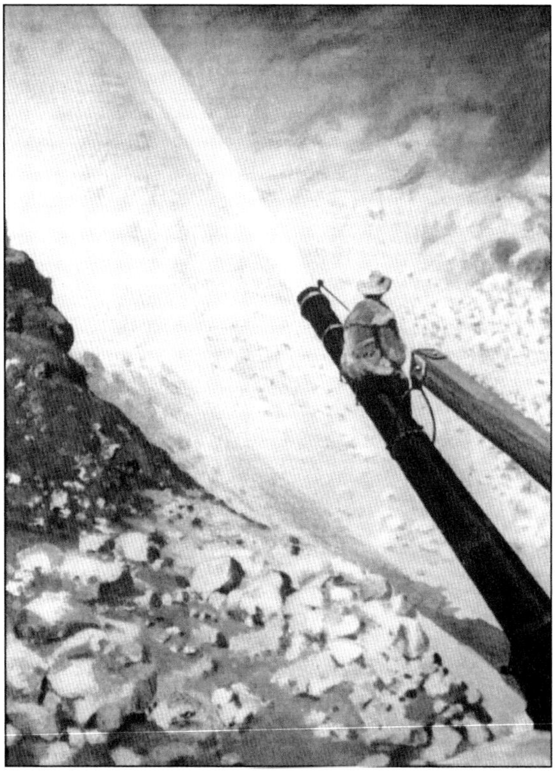

This Gold Flat image makes the hose seem tame with the man deftly seated upon the high-pressure nozzle. The water jet really gives a sense of the power applied to mountainside as it swipes it away; sitting on the monitor would have been quite a ride, with its constant vibrations. In 1852, Antoine Chabot invented the original monitor, consisting of a rawhide hose and a hand-carved wooden nozzle. Later, miners used hoses made from canvas with a nozzle made of iron. Crinoline replaced the canvas in the 1860s. (JLJ.)

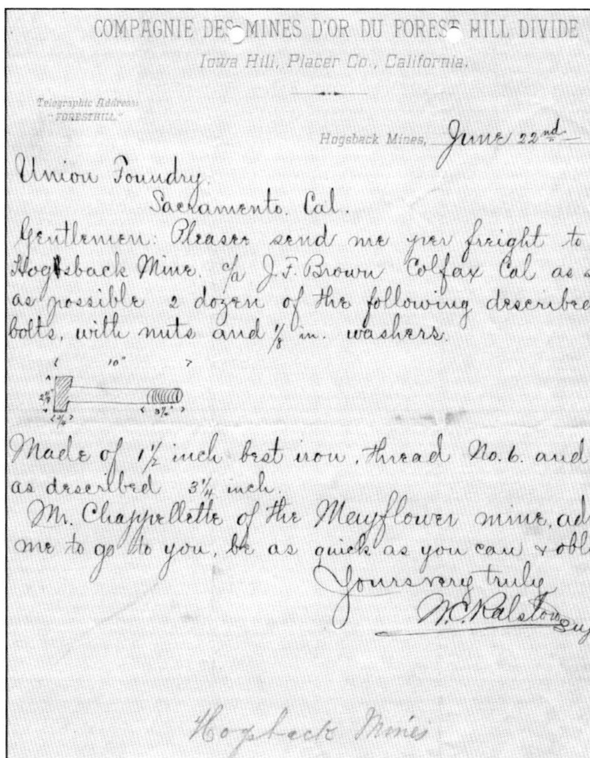

COMPAGNIE DES MINES D'OR DU FOREST HILL DIVIDE
Iowa Hill, Placer Co., California.

Telegraphic Address: "FORESTHILL."

Hogsback Mines, June 22nd

Union Foundry,
Sacramento. Cal.
Gentlemen: Please send me per freight to Hogsback Mine. c/a J.F. Brown Colfax Cal as s as possible 2 dozen of the following described bolts, with nuts and 1/8 in. washers.

Made of 1½ inch best iron, thread No. 6. and as described 3¼ inch.
Mr. Chappellette of the Mayflower mine, adv me to go to you, be as quick as you can & obli
Yours very truly
W.C. Ralston Sup

Hogsback Mines

In 1888, Companie des Mines D'or du Forest Hill Divide placed an order for construction supplies for Hogsback Mine in Iowa Hill. W.C. Ralston requested that the Union Foundry manufacture a given quantity of nuts, bolts, and washers. Ralston specifies his exact requirements, complete with drawings and dimensions. (AS.)

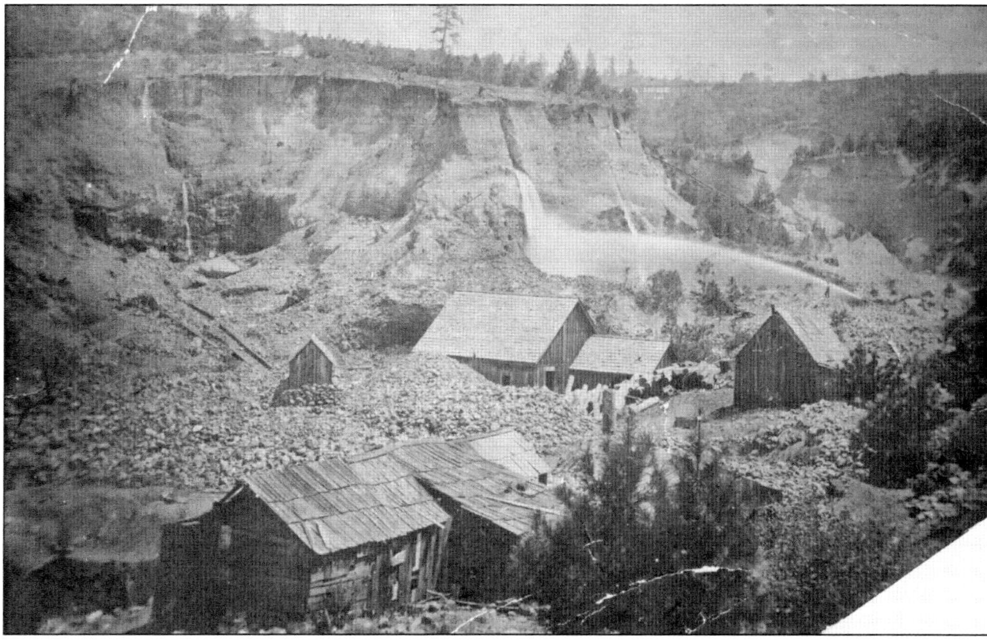

The title of this image, *Digs and Shacks*, aptly describes this environment and shows how close the operation's facilities and shacks were to the water-blasts and resulting waterfalls. Though photographs can be deceptive, some buildings look like they will soon be submerged in mounds of debris. The man-made ravine staggers the modern eye and consciousness. (JLJ.)

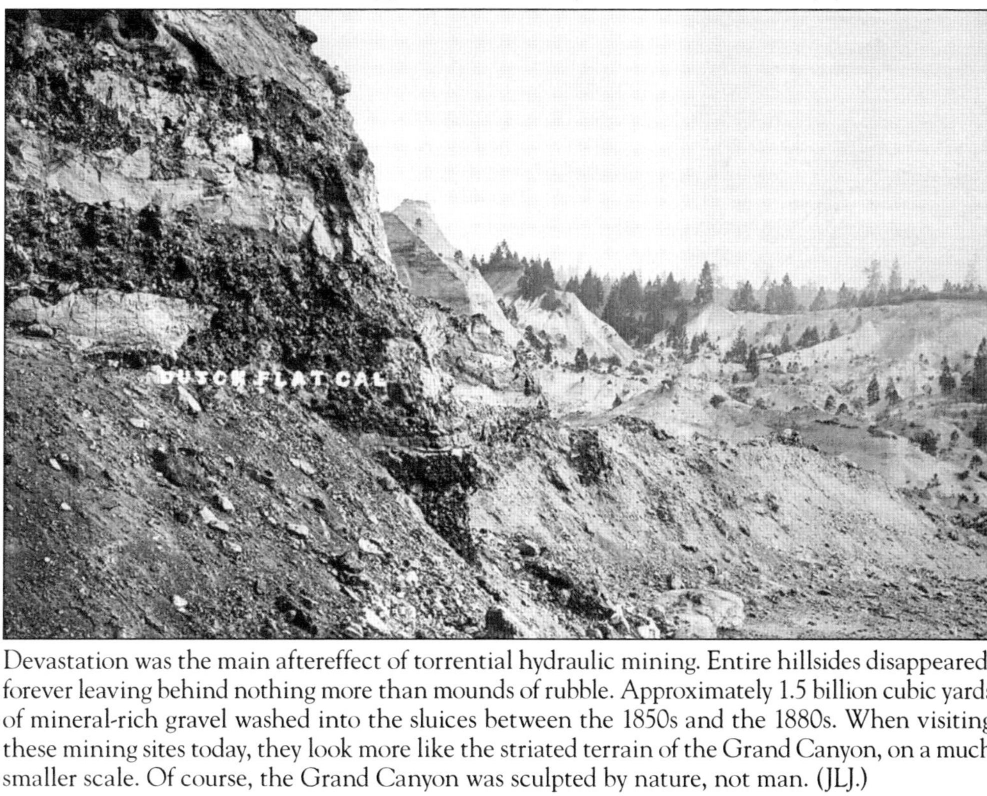

Devastation was the main aftereffect of torrential hydraulic mining. Entire hillsides disappeared, forever leaving behind nothing more than mounds of rubble. Approximately 1.5 billion cubic yards of mineral-rich gravel washed into the sluices between the 1850s and the 1880s. When visiting these mining sites today, they look more like the striated terrain of the Grand Canyon, on a much smaller scale. Of course, the Grand Canyon was sculpted by nature, not man. (JLJ.)

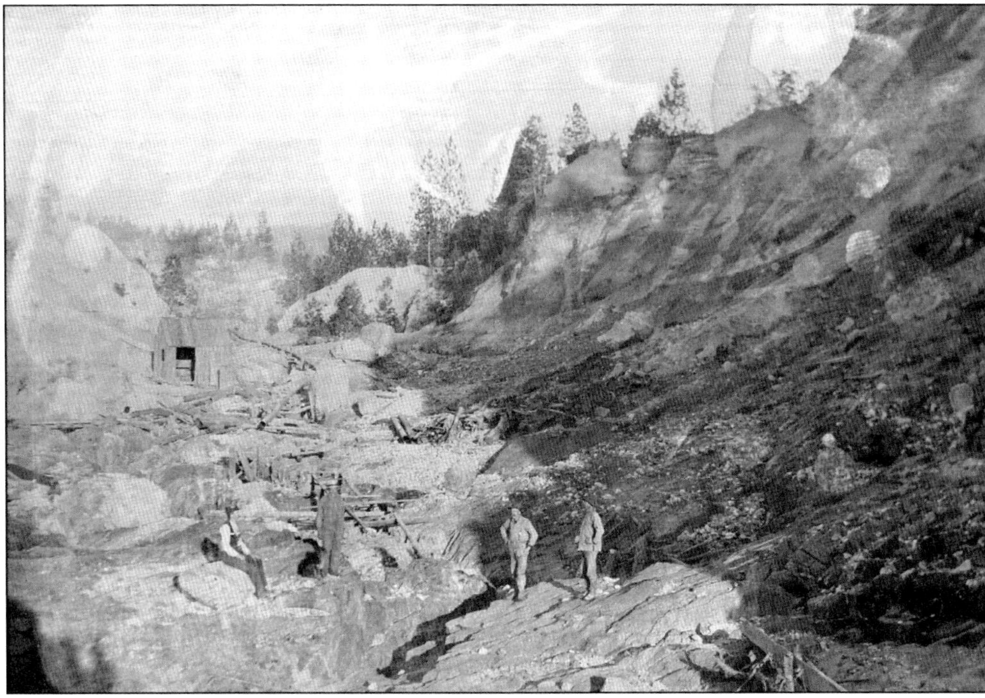

The new "scenery" at Gold Run (above) shows what men can do when they have gold fever. Forested hills still decorated the background, but they would soon disappear to meet railroad construction and underground mining demands. Through the years, the foothills have been reforested. (JLJ.)

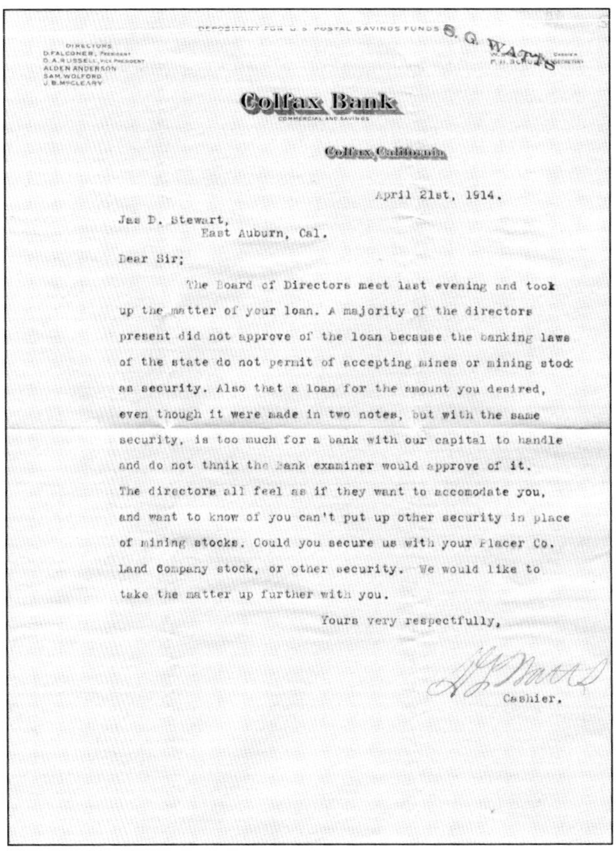

DEPOSITARY FOR U. S. POSTAL SAVINGS FUNDS

S. G. WATTS

DIRECTORS
D. FALCONER, PRESIDENT
D. A. RUSSELL, VICE PRESIDENT
ALDEN ANDERSON
SAM. WOLFORD
J. B. McCLEARY

F. H. SCRUTS, SECRETARY

Colfax Bank
COMMERCIAL AND SAVINGS

Colfax, California

April 21st, 1914.

Jas D. Stewart,
East Auburn, Cal.

Dear Sir;

The Board of Directors meet last evening and took up the matter of your loan. A majority of the directors present did not approve of the loan because the banking laws of the state do not permit of accepting mines or mining stock as security. Also that a loan for the amount you desired, even though it were made in two notes, but with the same security, is too much for a bank with our capital to handle and do not thnik the Bank examiner would approve of it. The directors all feel as if they want to accomodate you, and want to know of you can't put up other security in place of mining stocks. Could you secure us with your Placer Co. Land Company stock, or other security. We would like to take the matter up further with you.

Yours very respectfully,

H. G. Watts
Cashier.

Jas D. Stewart sought financing for his mining, but the Colfax Bank rejected his request in this letter dated April 21, 1914. With the gold standard set at \$20.67 per ounce, the board of directors ruled that neither his mine nor his gold-mining stock offered sufficient collateral required by the banking laws in California. (AS.)

Three

RAILROADING

During the Gold Rush, three primary routes brought prospectors to California: from Asia across the Pacific Ocean; from Europe around South America's Cape Horn to San Francisco; and from Europe to the Isthmus of Panama, across land, and then north to San Francisco. Hardy souls trekked west over ranges of mountains, across prairies and deserts in Conestoga wagons, on stagecoaches, on horseback, and even on foot. Some gave up along the way and settled down at an intermediate location. Some turned back, and some did not live through the experience.

The cross-country trip in the mid-1800s took a long time no matter the choice of transportation. Forward-thinking Theodore Judah, a railroad engineer from New York, traveled between Washington, DC, and California to assess possibilities for a transcontinental railroad. He recruited four investors, who became known as the Big Four—Leland Stanford, Collis P. Huntington, Charles Crocker, and Mark Hopkins—to support his proposed railroad. In 1862, Congress passed the Pacific Railroad Act, awarding contracts to both Central Pacific and Union Pacific. President Lincoln concurred largely because the Civil War had begun and he envisioned the railroad transporting Union troops and gold.

Construction of the new railroad posed a massive challenge. The Union Pacific Railroad Company crews headed west from the Mississippi River, and the Central Pacific Railroad team started from Sacramento heading east. They met at Promontory, Utah, where they drove the Golden Spike. Many consider the struggle to lay rails over the rugged Sierra Nevadas, where the elevation rises 7,000 feet in just 23 miles, to be the most impressive achievement.

Nevada City and Grass Valley needed a way to tie into the new transcontinental railroad running through Colfax, and a narrow gauge railroad was completed in 1876. It gave Nevada County ready access to the world. For 66 years, the Nevada County Narrow Gauge Railroad proved invaluable for freight and passenger use. In 1942, the company repainted the cars dark green and relettered them "Navy Supply Depot, Pearl Harbor" for redeployment in World War II.

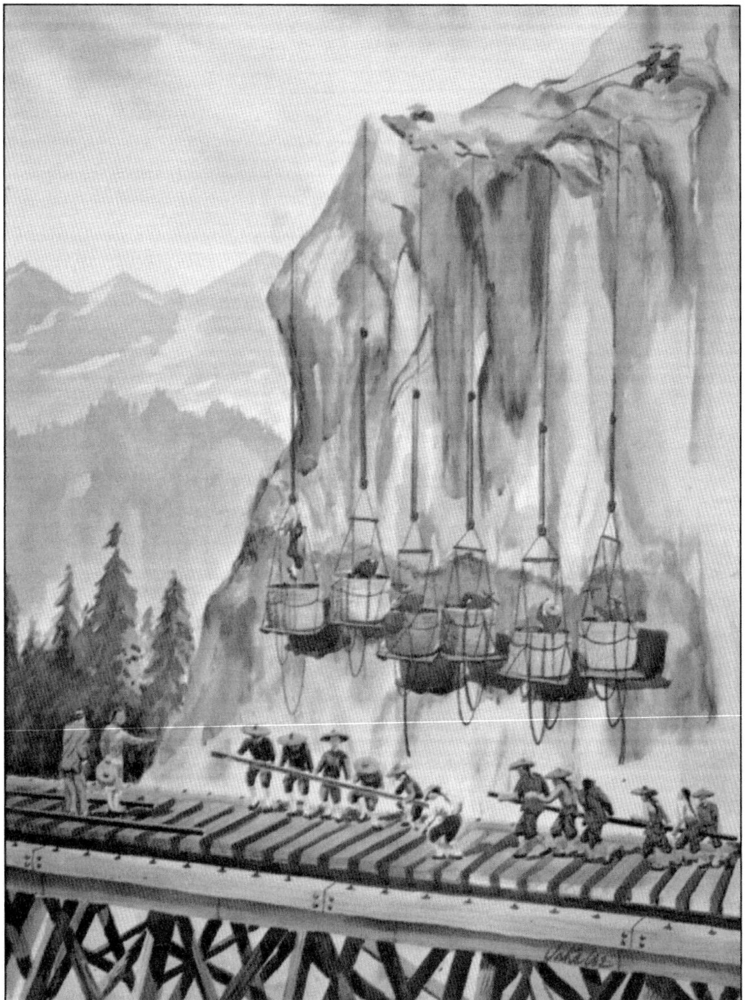

In the Sierra. Thousands of Chinese laborers built the Central Pacific tracks which joined the Union Pacific to span America in 1869.

美國最初鐵路工程均由華人建築

© 1963 J. K.

This painted postcard shows a famous rendition of Chinese workers being lowered down mountain cliffs in bosun chairs while working on the transcontinental railroad in the Sierras. Jack E. Duncan refuted the theory of bosun chairs in his 2005 book, *A Study of Cape Horn Construction on the Central Pacific Railroad 1865–1866*. His research over three years concluded that the slopes were insufficiently steep to warrant baskets. He did not claim that workers were not working in precarious situations on the mountainside, but decided that ropes were probably used without bosuns. Not all historians concur with Duncan, and no written accounts document that time. Workers did not use machinery but wielded picks by hand then planted the explosive black powder. This Cape Horn, located one mile east of Long Ridge Ravine, was reportedly named after the other famous Cape Horn, which challenges ship captains with unpredictable seas when rounding South America. (CC.)

Cape Horn, approximately three miles east of Colfax by rail, is one of the most challenging sections of the Sierras conquered by the Central Pacific Railroad during the building of the transcontinental railroad. Most historians associate Cape Horn with hard labor and the industrious Chinese workers who made it possible for trains to negotiate the mountain. (AS.)

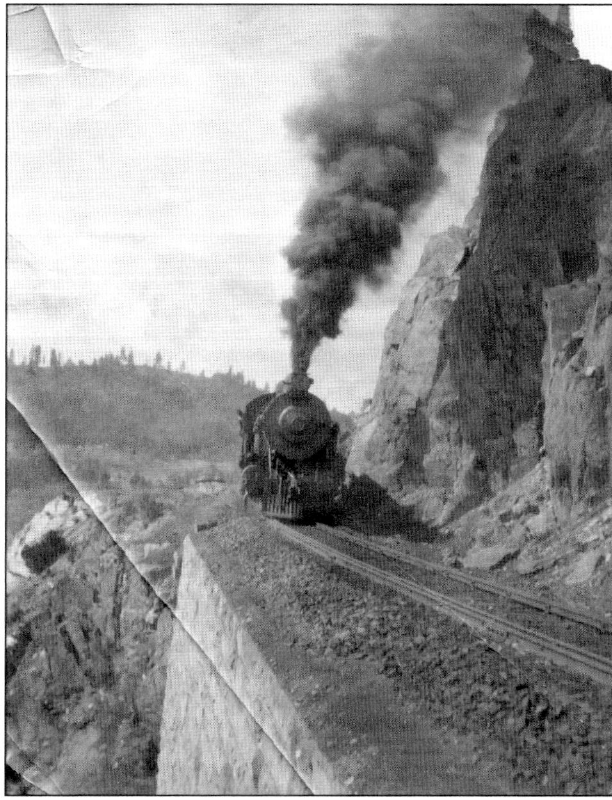

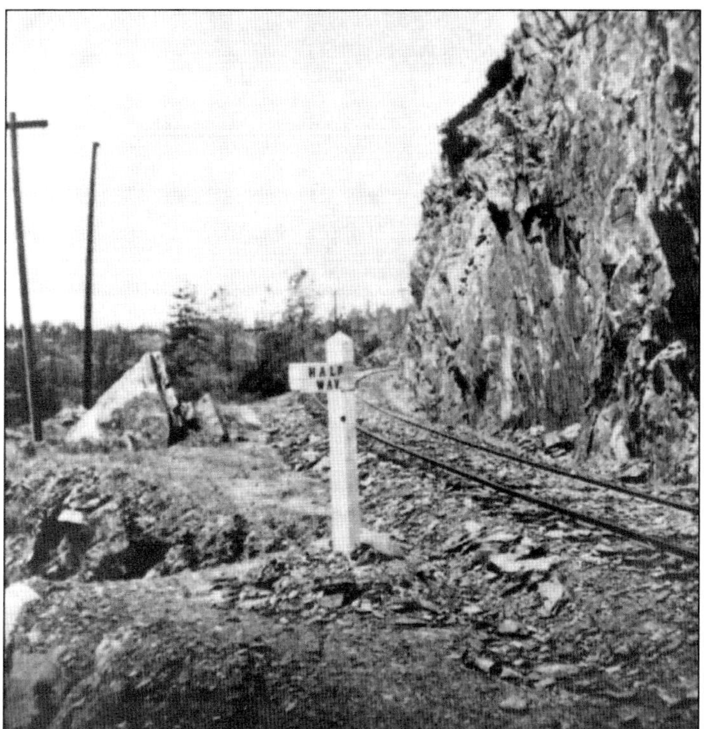

The Central Pacific posted a sign marking the halfway point around Cape Horn. Looking at the terrain, imagine the thrill of the railroad construction workers when they realized they had picked and blasted their way around half of this barrier. Men also carved tunnels through the mountains. In December 1914, they bored two tunnels, one 1,300 feet long and the other 400 feet long, through the Sierras at Cape Horn to enable double tracking. This arrangement let the railroad use each track exclusively in one direction. (JLJ.)

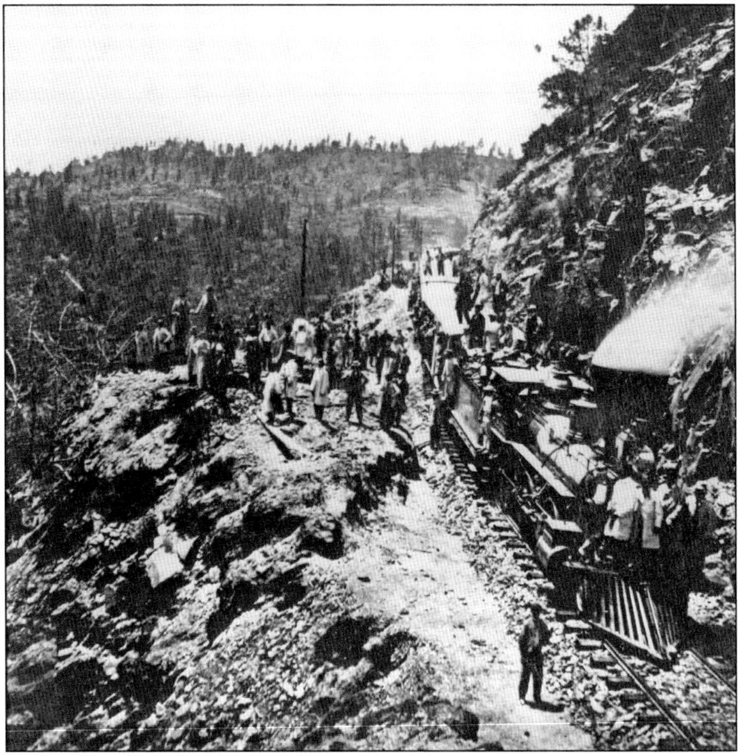

Passengers disembark the train temporarily at Cape Horn to take in the breathtaking view of the mountains and the river below. This early shot emphasizes the rocky, rough-hewn surroundings at Cape Horn. The ledge looks down to the North Fork of the American River more than 1,000 feet below. Many climbed atop the railcars for a better view. (JLJ.)

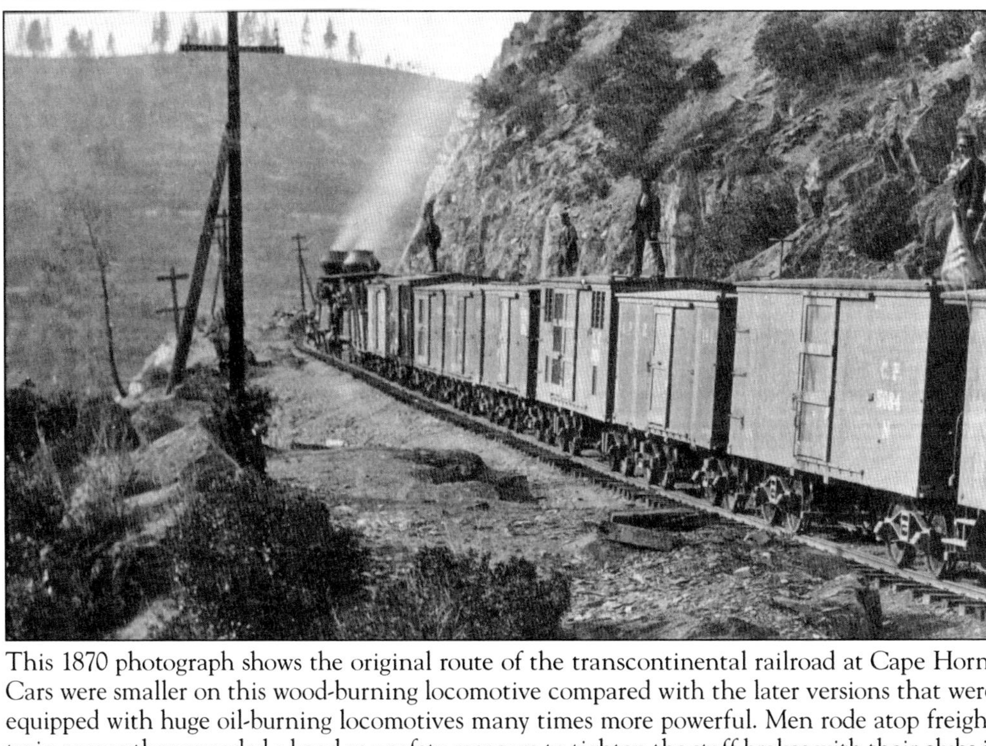

This 1870 photograph shows the original route of the transcontinental railroad at Cape Horn. Cars were smaller on this wood-burning locomotive compared with the later versions that were equipped with huge oil-burning locomotives many times more powerful. Men rode atop freight train cars as they rounded a bend as a safety measure to tighten the staff-brakes with their clubs if the engineer whistled. The process did not always prove reliable, and accidents did occur. (AS.)

A lady, her two daughters, and their dog stand on the cross ties of the railroad tracks staring out to the ravine below. This image further illustrates the curiosity of the visitors who disembarked the train for the view. The monstrous retaining wall, made from granite, was added for strength and stability. The original wall, built in 1895, was made from "dry rubble," meaning that the rocks were stacked but not mortared in place. Mortar was added later to accommodate larger, heavier trains. (JLJ.)

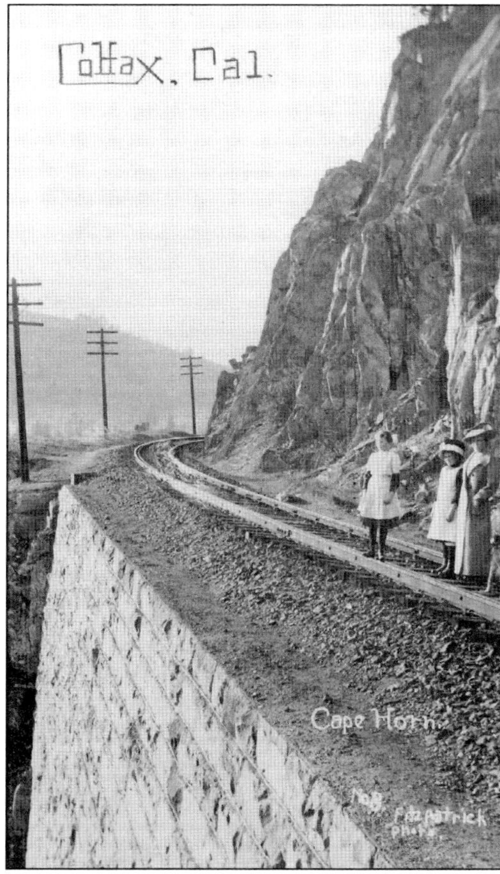

The close-up photograph below shows an overland passenger train stopped at Cape Horn in 1870. Riders stretch their legs for a spell and take in the view. This pause for reflection became a common practice for engineers, to the certain enjoyment of their passengers. It appears that these riders were predominantly men; their beards, hats, and long jackets are representative of the era. (JLJ.)

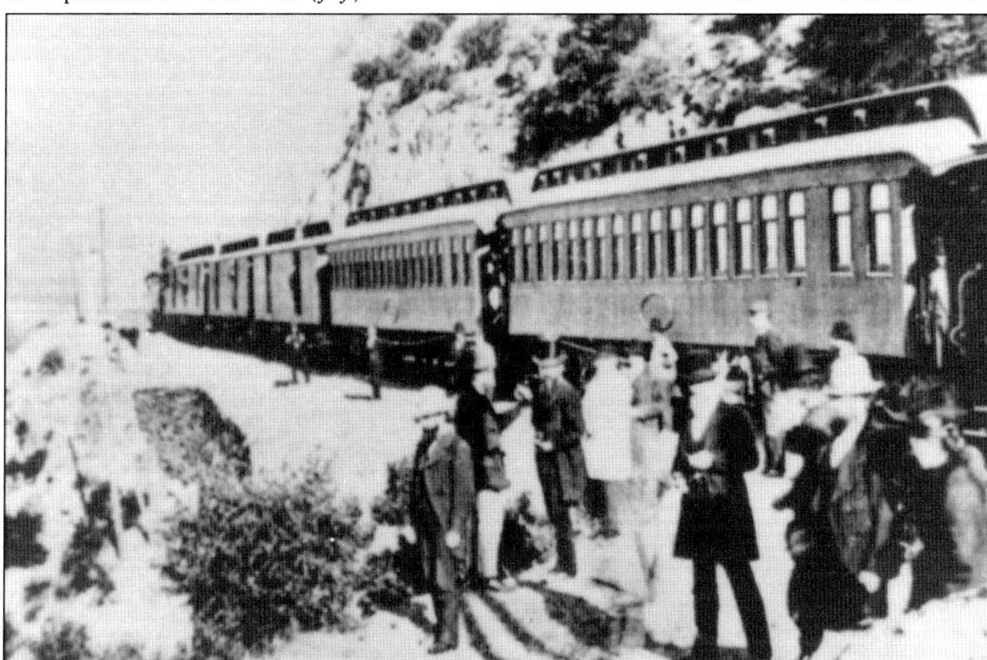

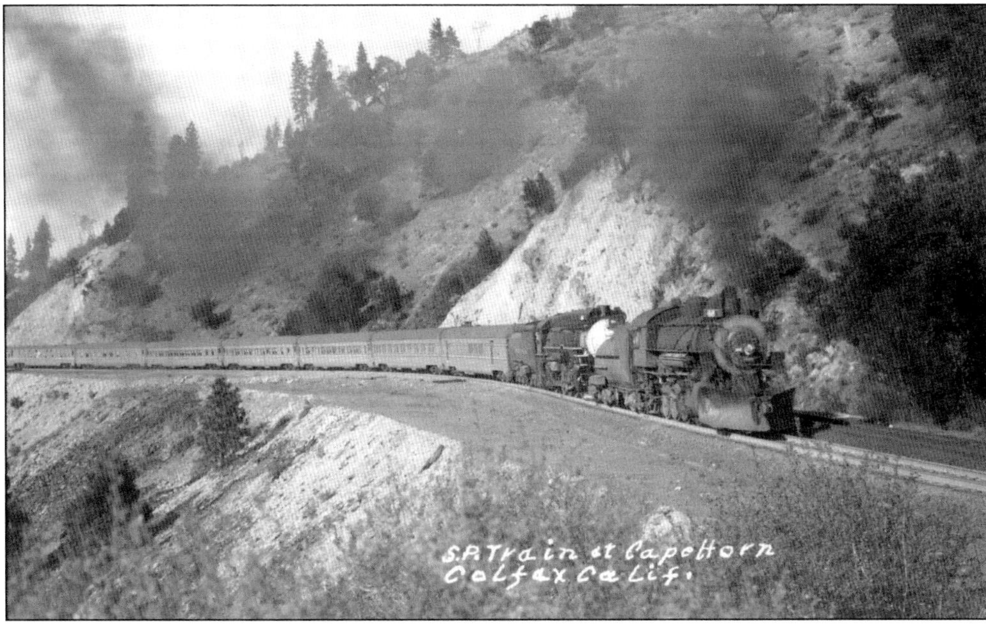

In this image, note how clean and smooth the surroundings are compared with the two previous photographs, where the tracks were surrounded by rubble, making it difficult to see how the railroad ties held the rails in place. This train has two engine units pulling it. This arrangement provided the extra power required for steep grades. (AS.)

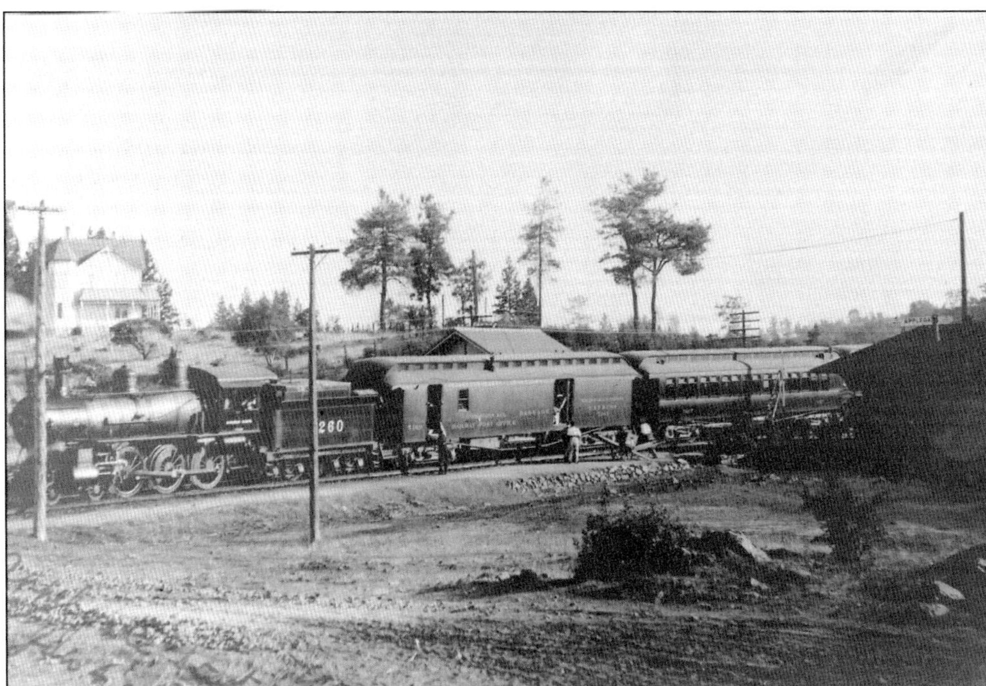

In the foreground, a locomotive rests in the town of Colfax as men load or unload the baggage car. The remaining cars of the train appear to be passenger cars. In the distance up on the hill, a Victorian house overlooks the spare surroundings. (JLJ.)

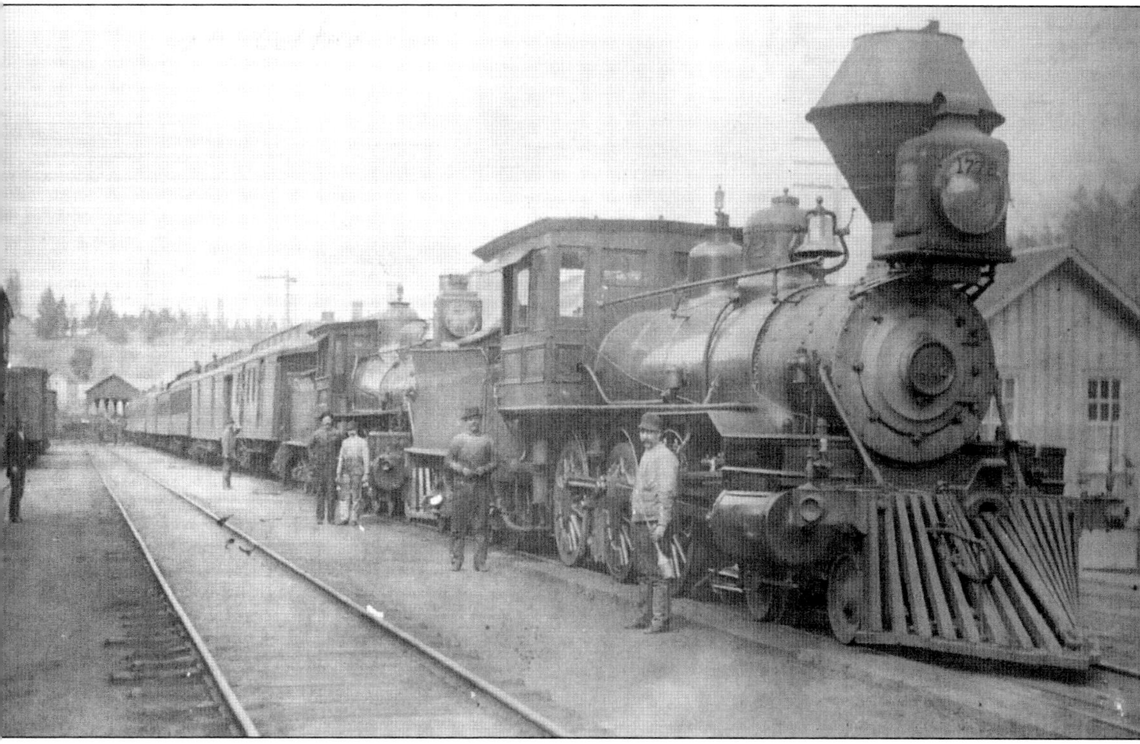

Central Pacific Railroad workers pose in front of two rear-cab locomotives. The front one, No. 1772, is a wood-burning model, evidenced by the large coned smokestack at the top of the engine. Engine No. 1726 (rear) is harder to distinguish in this photograph. In the eastern United States, locomotives burned coal, but out west, coal was unavailable and wood was initially plentiful. This postcard is dated January 1, 1891, on the back of the postcard. (PCM.)

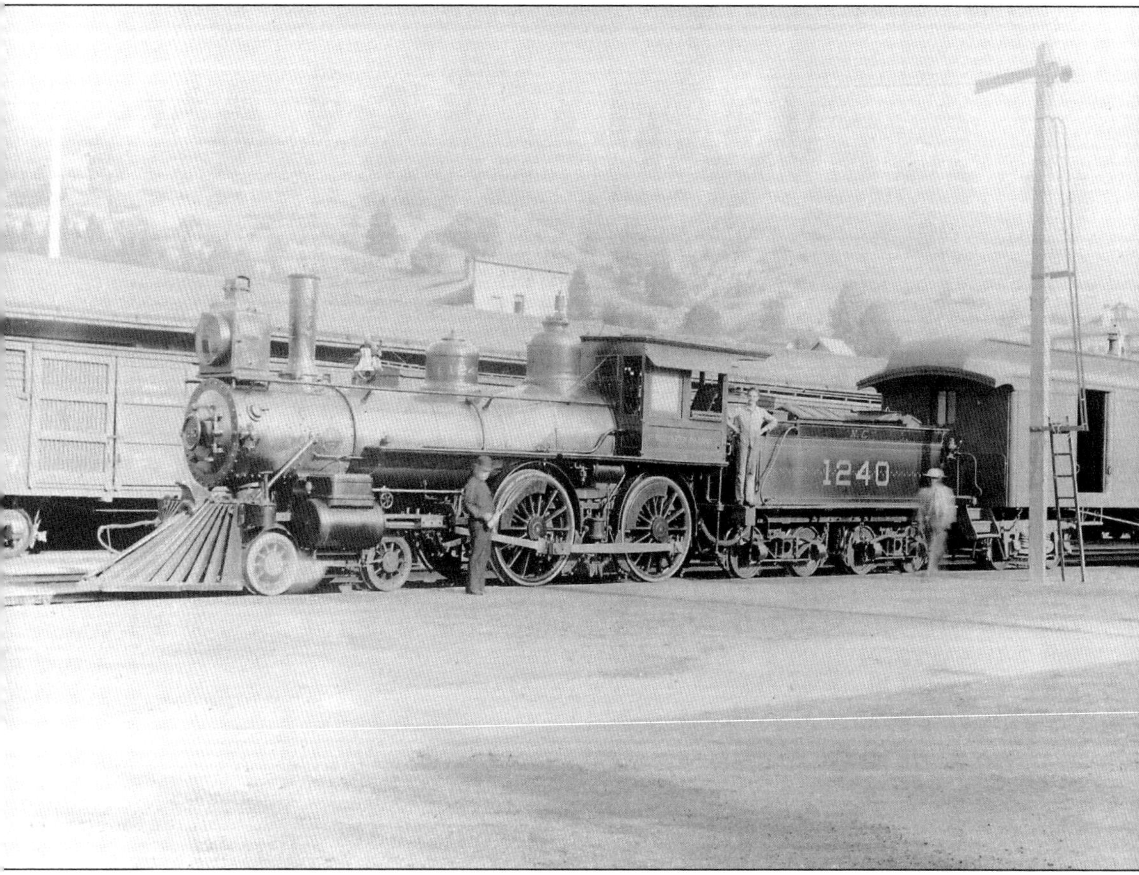

Locomotive No. 1240, a steam engine, is stopped at Colfax. Note the design difference in the chimney on the top of the engine. It is cylindrical and narrower than the wood-burning model. The cowcatchers in this region were valuable for pushing aside snow on the higher tracks during the winter. (JLJ.)

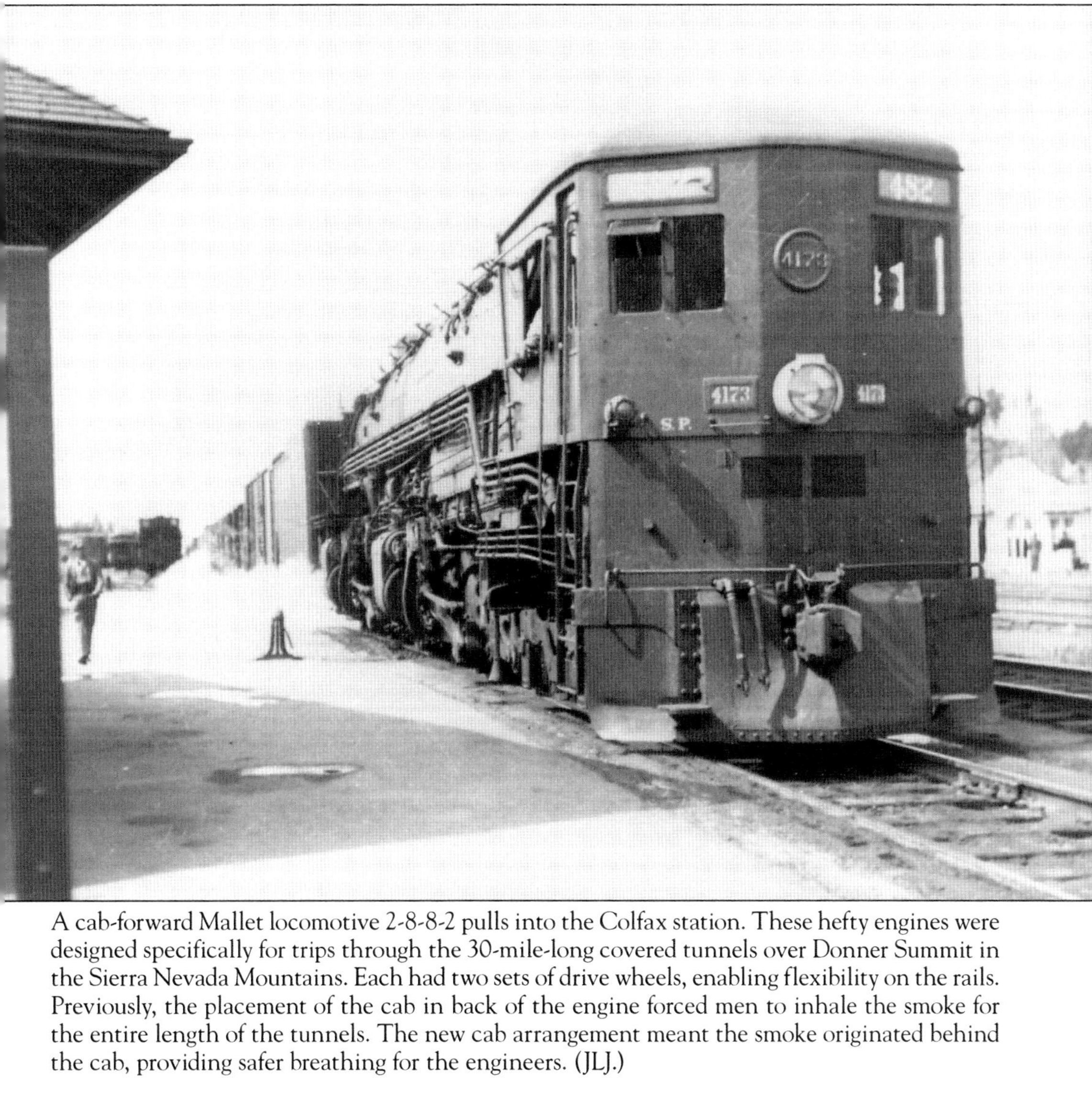

A cab-forward Mallet locomotive 2-8-8-2 pulls into the Colfax station. These hefty engines were designed specifically for trips through the 30-mile-long covered tunnels over Donner Summit in the Sierra Nevada Mountains. Each had two sets of drive wheels, enabling flexibility on the rails. Previously, the placement of the cab in back of the engine forced men to inhale the smoke for the entire length of the tunnels. The new cab arrangement meant the smoke originated behind the cab, providing safer breathing for the engineers. (JLJ.)

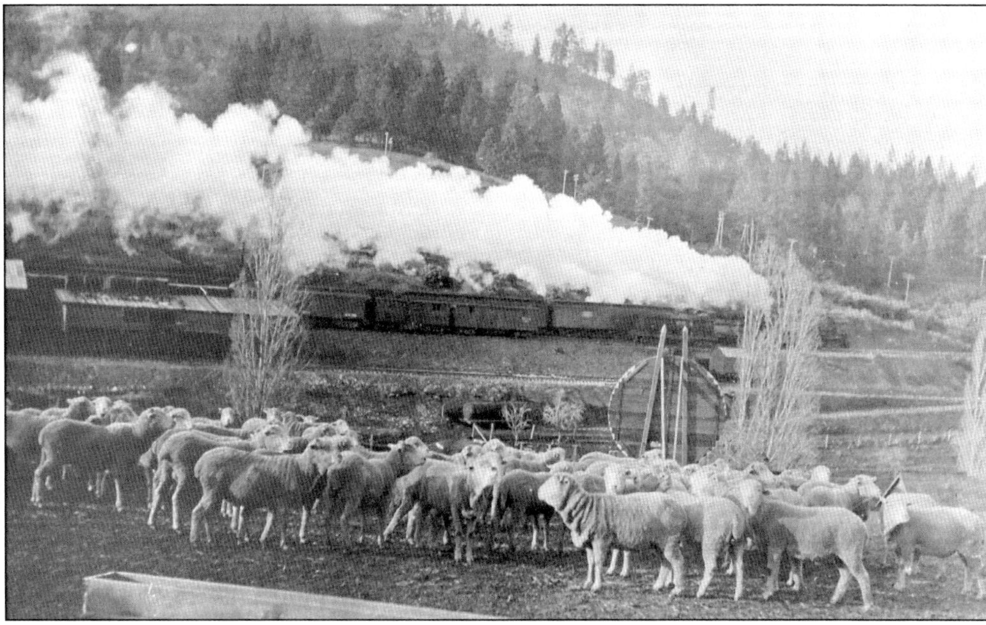

The steam billows back from the Colfax train as it pulls out of town, almost masking the hillside view. In 1912, helper engines based in Colfax provided extra pull over the summit. In the foreground, sheep huddle tightly together near their trough as they watch the iron horse rumble by. (JLJ.)

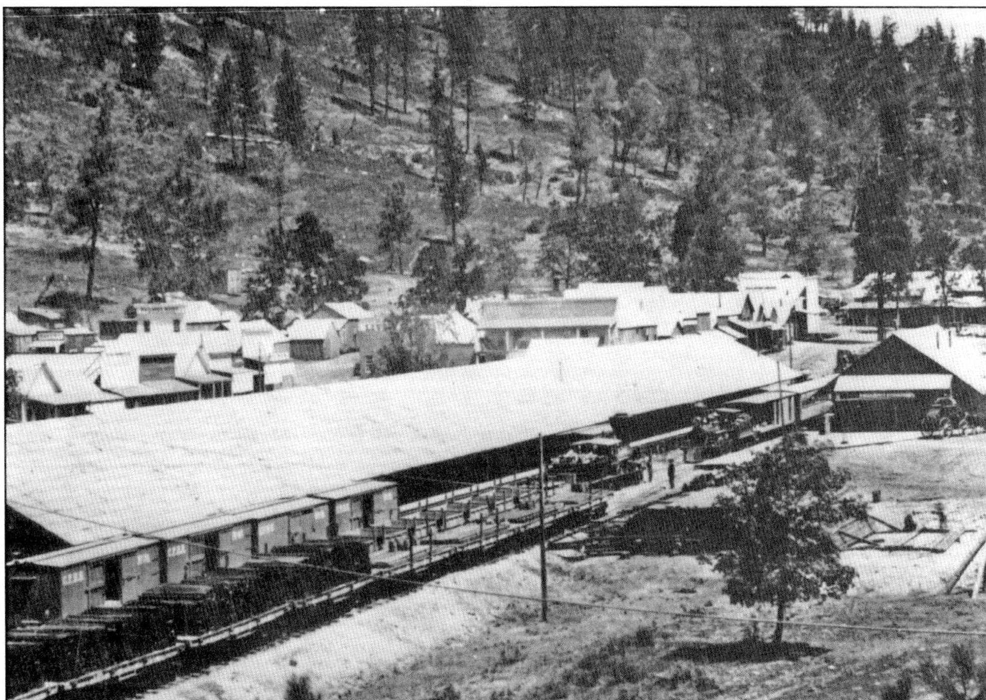

Seen here are the two original depots in Colfax parallel to the early town. A stagecoach—a typical conveyance for travelers to outlying communities—waits at center right. Supplies transported by rail transferred from the cars to wagons or pack trains for delivery. (CHM.)

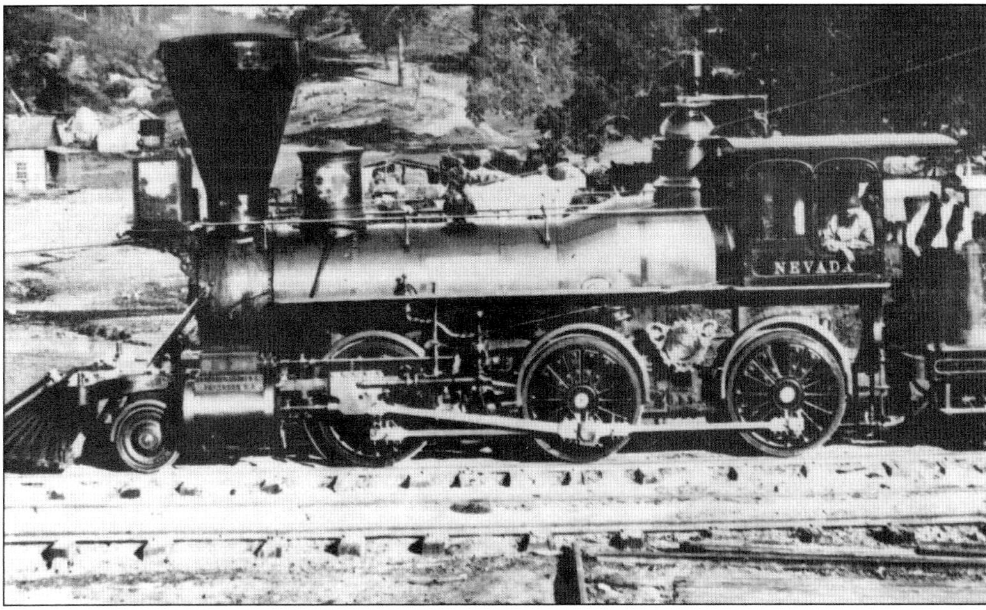

This close-up view of the locomotive *Nevada* shows the engineer peering out of the rear-cab. Behind the train, the scenery reflects the stripped landscape resulting from the rush to fortune in the early days of California. (CHM.)

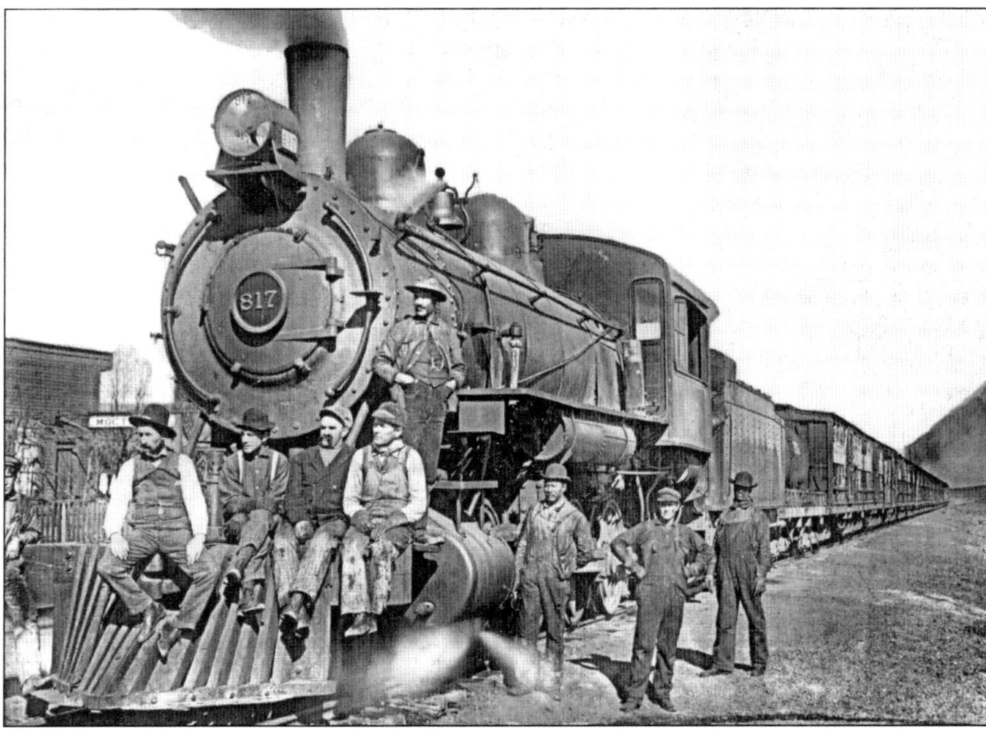

Locomotive 817, with its rear-cab design, sits on the tracks almost as proudly as the workers beside it and those posing to the right of the engine. The ones standing seem to be the hardest workers, if their dirty coveralls are a valid indication. All the workers still made sure to don their hats for the photograph. (CHM.)

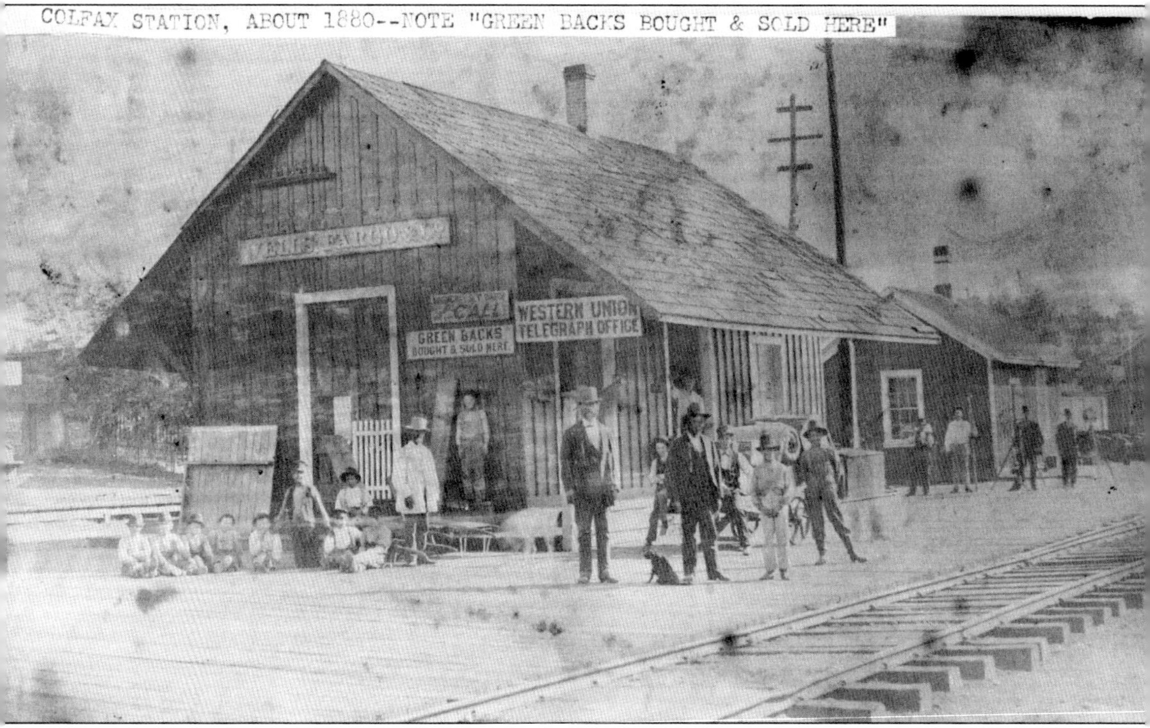

In 1880, men and boys, all in hats, dawdle in front of the Wells Fargo building, which advertises "Green backs bought & sold here." This served as an assay office for miners to exchange their gold into paper money. The railroad transported the gold to the city banks. On February 25, 1862, Congress approved the issuance of paper money under the Legal Tender Act. The US Treasury paid Union soldiers with the new greenback bills instead of strictly gold and silver. (PCM.)

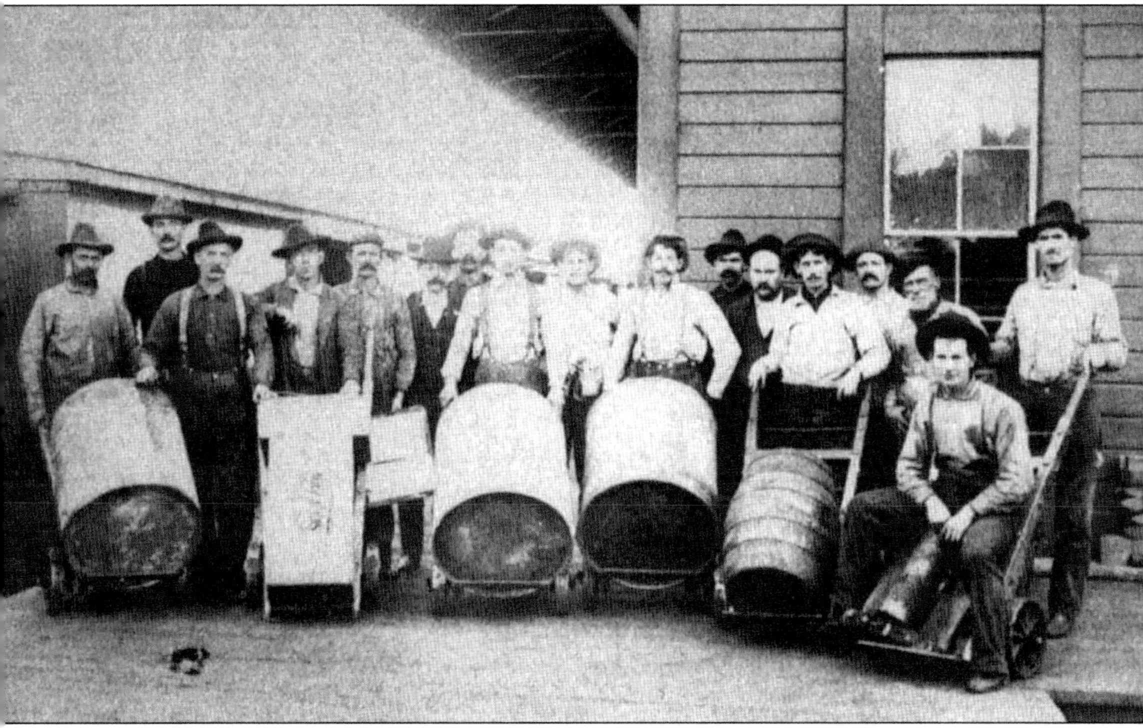

Railroad depot freight workers pose next to the station in November 1898. A list of employee names reads, from left to right, W.A. Bruton (from Tennessee), ? Burges (brakeman), Bob Russell, B. Harris (boss), Frank Kuenzly, J.A. Dyer (Sam), M. Lobner (agent), A. Burges (conductor), Jas Thomas (straight edge), W. Clark, J.L. Butler (Pills), R.D. Brown (Robinson Crusoe), L. Sliuman (assistant agent), C. Clark (Pat), B.W.S. Perkins, G. Moret (Toppy), ? Hicks (Friday), and E. Murray (Klondike, truck man). Responsibilities or nicknames of the individuals are noted in parentheses. (CHM.)

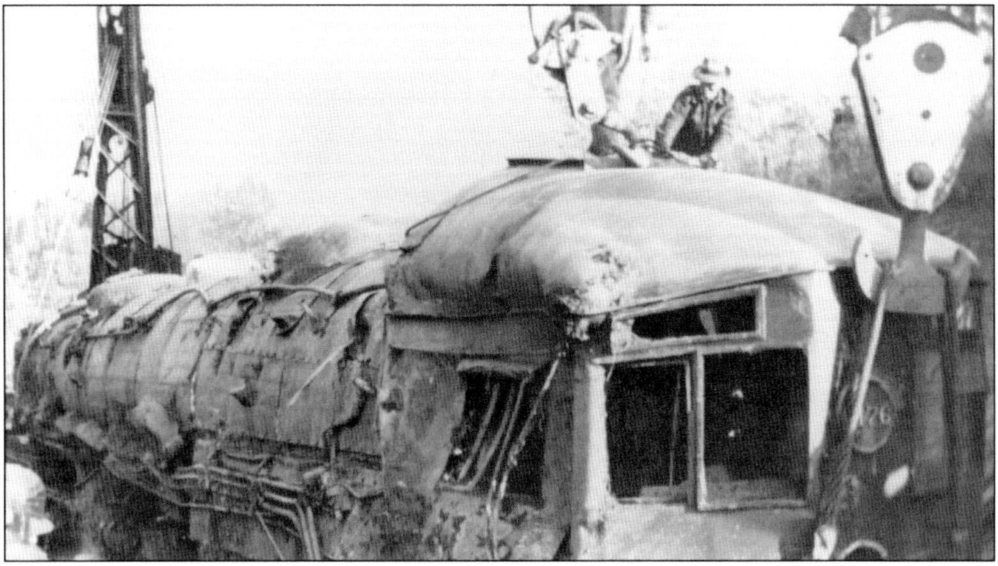

Working for the railroad, either in its construction or later in its operation, posed dangers. This shows the westward-bound, cab-forward engine called the *Challenger.* It chugged along only 1.5 miles out of Colfax when a wreck occurred on November 8, 1944, at 5:10 a.m. Another famous wreck occurred in the late 1880s near Cape Horn when a cattle car and several others heaved over the edge when rails separated. The injured cows panicked, resulting in a stampede through town. (PCM.)

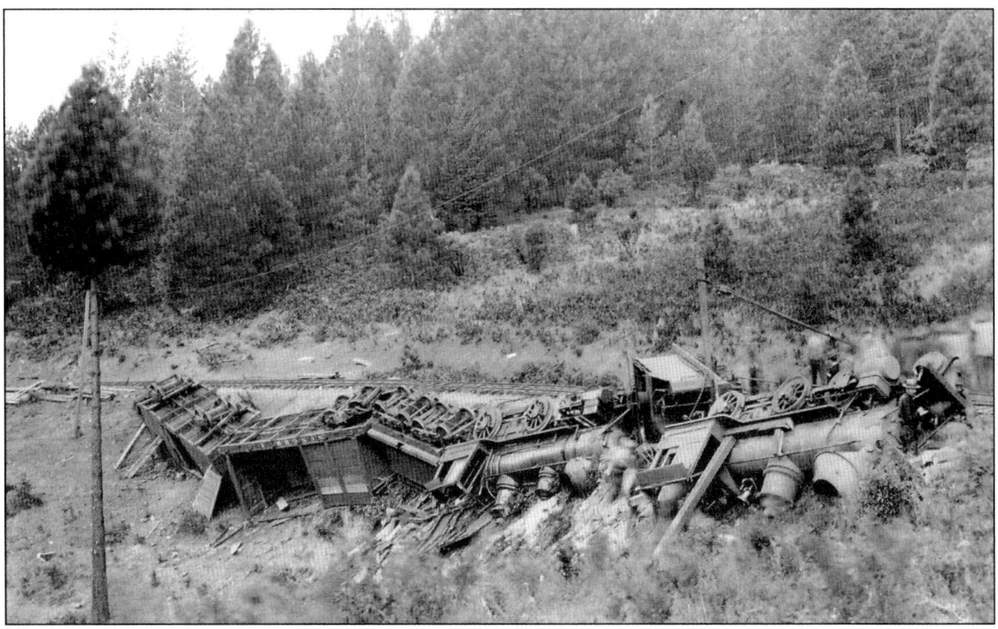

Each year, a circus came west. The company transported its equipment and animals via the transcontinental railroad. A disaster occurred in 1893 when horses shifted to the side in one car, causing the train to tip and jump the tracks, expelling a number of animals. Monkeys had to be coaxed down from the trees. Recovered equipment and animals went on to Nevada City by flatcar to perform, but the elephants walked. On the next trip, the circus only put dogs and ponies on the railroad. (CC.)

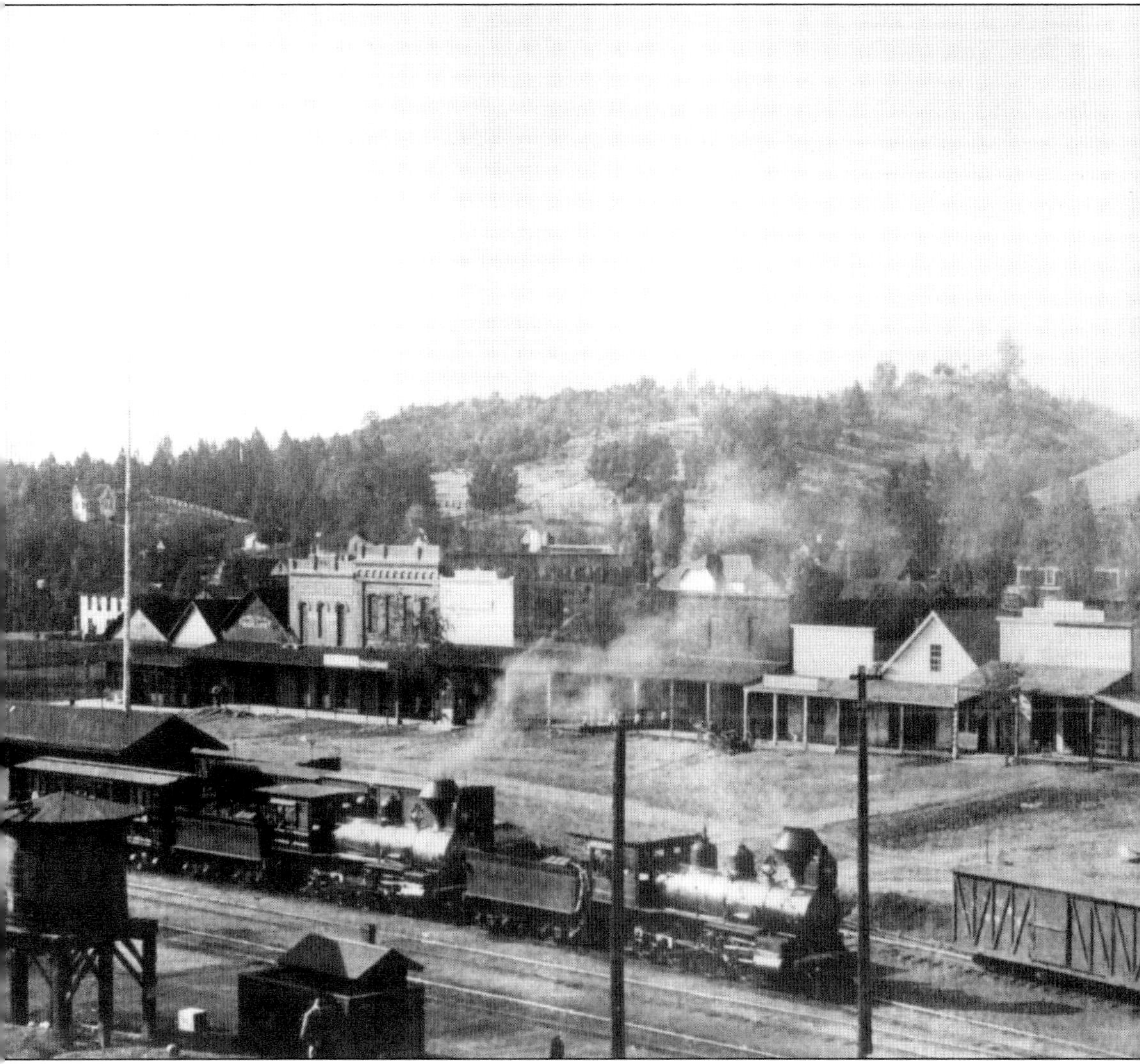

This 1893 picture shows a train pulling through the Colfax station with two engines attached in front to add extra power over the steep terrain. Both engines are wood-burning models. A large water tank is mounted on supports in the left foreground. It served the steam engines. This view shows the center of town on Main Street. The large brick buildings were constructed after fires became a significant threat to wooden buildings. (AS.)

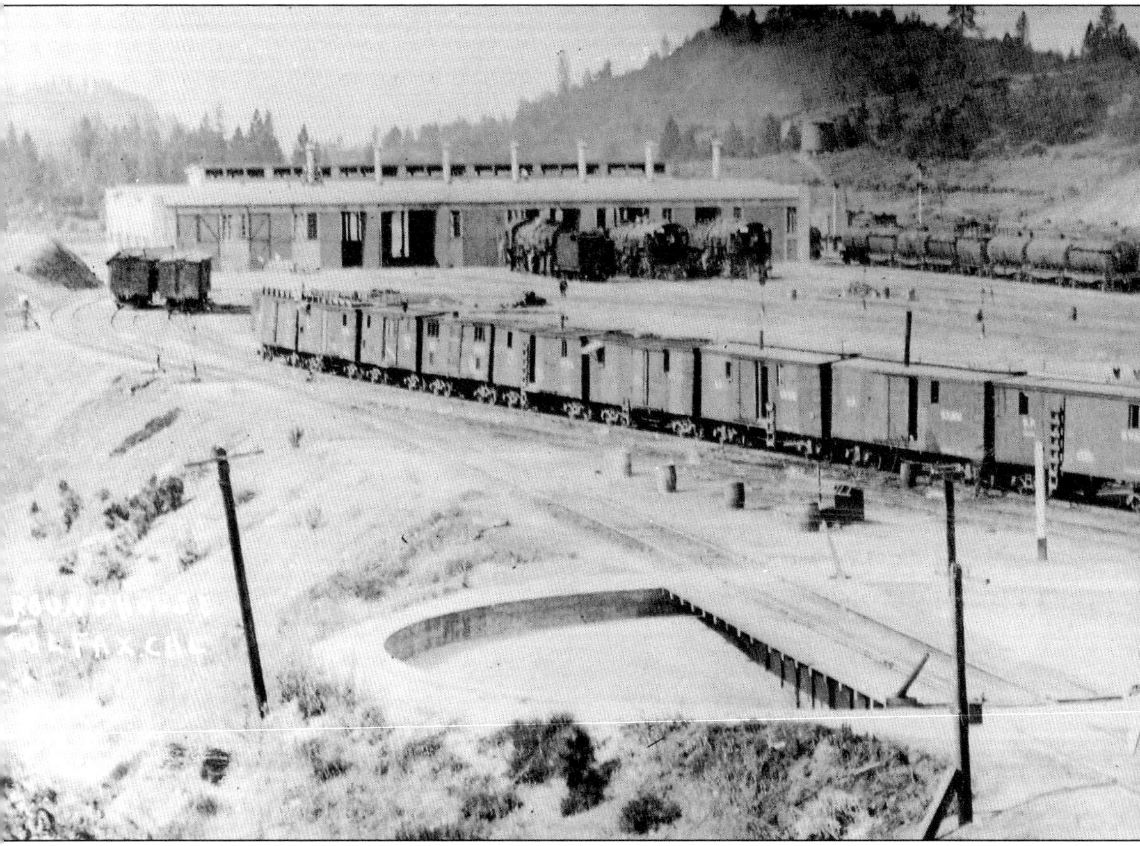

The 65-foot turntable in the foreground, which revolved the locomotives, was located one mile down the tracks from the depot in Colfax. It was torn down in 1905. Early steam engines were generally operated moving forward and needed rotating around to travel in the opposite direction. In 1912, the railroad built a wye track when it erected the engine house. The wye consisted of a triangular junction with a switch and turned the engines around in that way. (JLJ.)

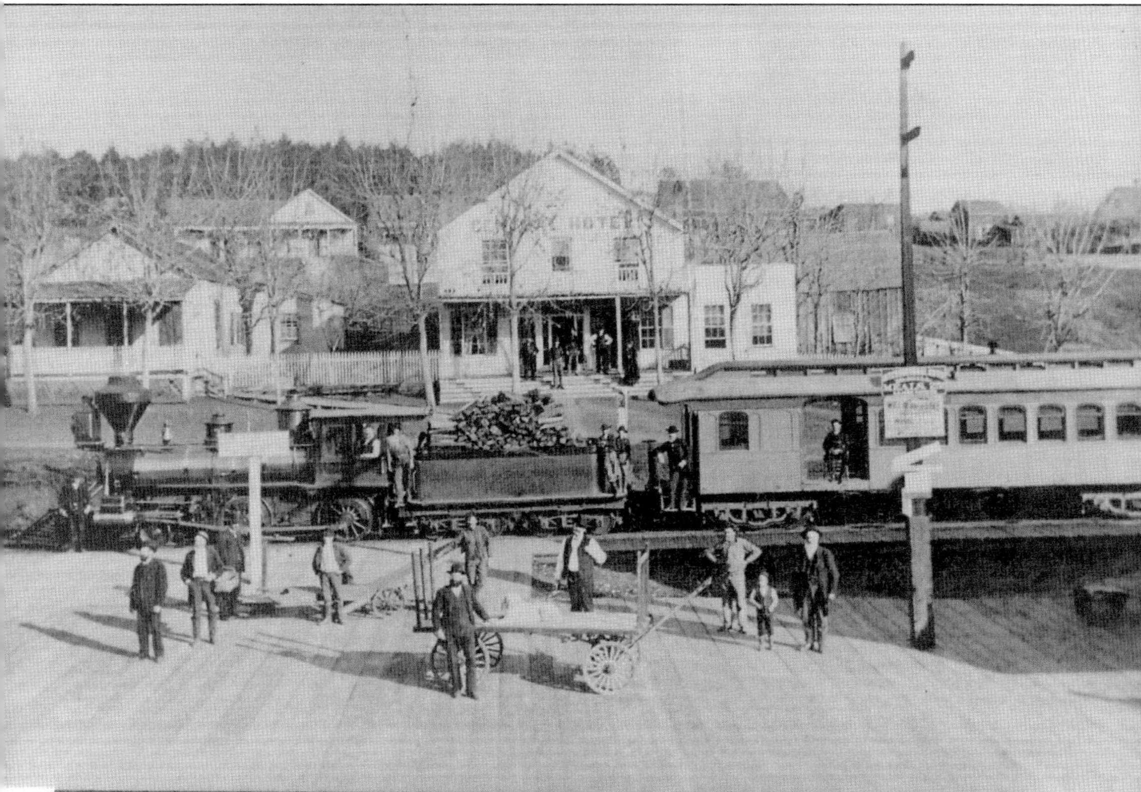

Narrow Gauge Train at Colfax

The Nevada County Narrow Gauge Railroad (NCNGRR) was established in 1876 to provide service from Nevada City to Colfax. The NCNGRR acquired the local nickname "Never Come Never Go Railroad," even though it had a reputation for being punctual. This narrow-gauge train sits opposite the Central Hotel in Colfax. (JLJ.)

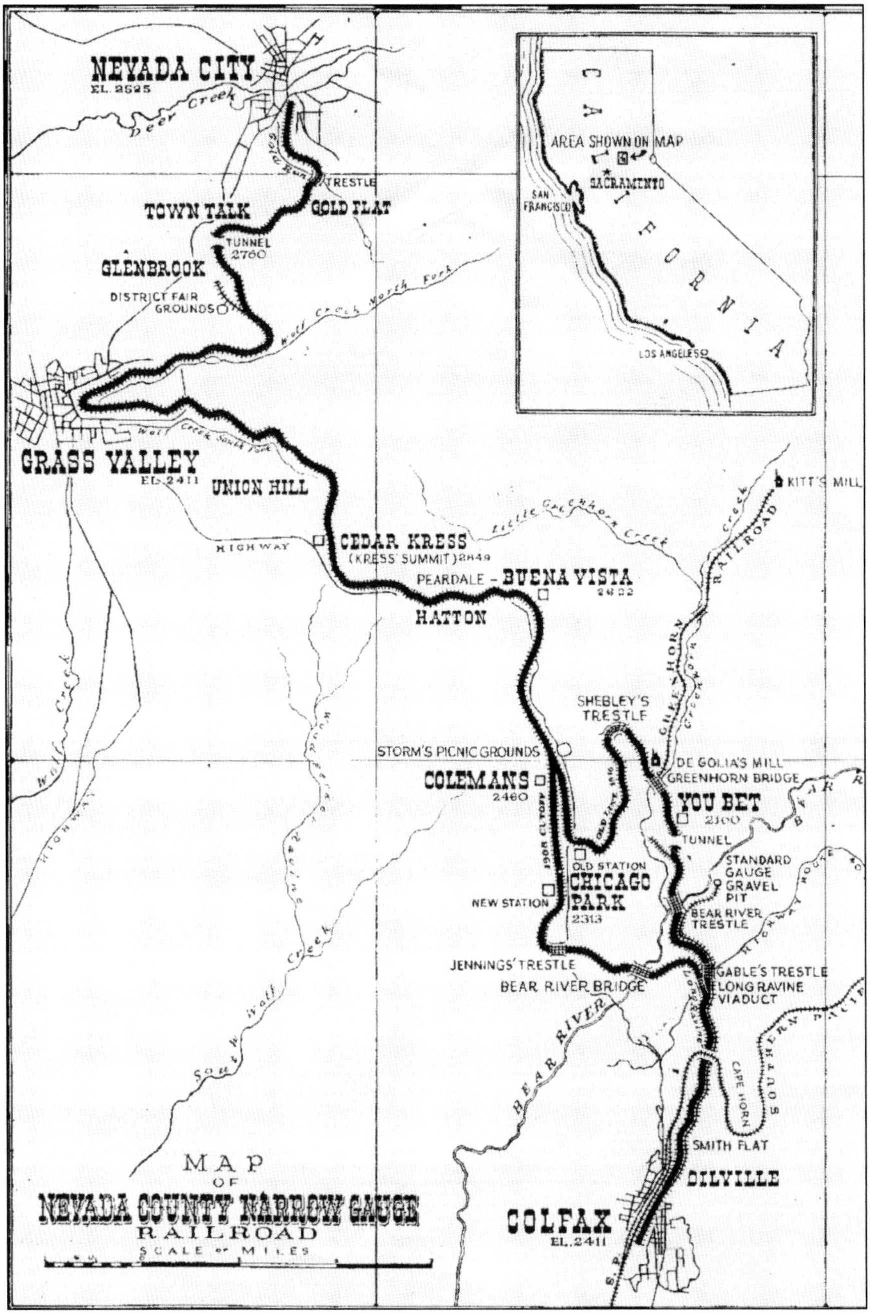

This Nevada County Narrow Gauge Railroad map indicates the entire route of the trains between the depot in Nevada City and the terminal at Colfax, where freight and passengers could connect to the transcontinental railroad. The tracks included two bridges, two tunnels—named Town Talk and You Bet—multiple trestles, and numerous twists and turns. After departing Colfax, passengers would cross the Bear River into Nevada County. There were multiple stops along the way—at You Bet, Chicago Park, Buena Vista, and Grass Valley—before reaching Nevada City. Famous passengers included Emma Nevada, Great Britain's Prince Albert, and US presidents Ulysses S. Grant and Theodore Roosevelt. (CHM.)

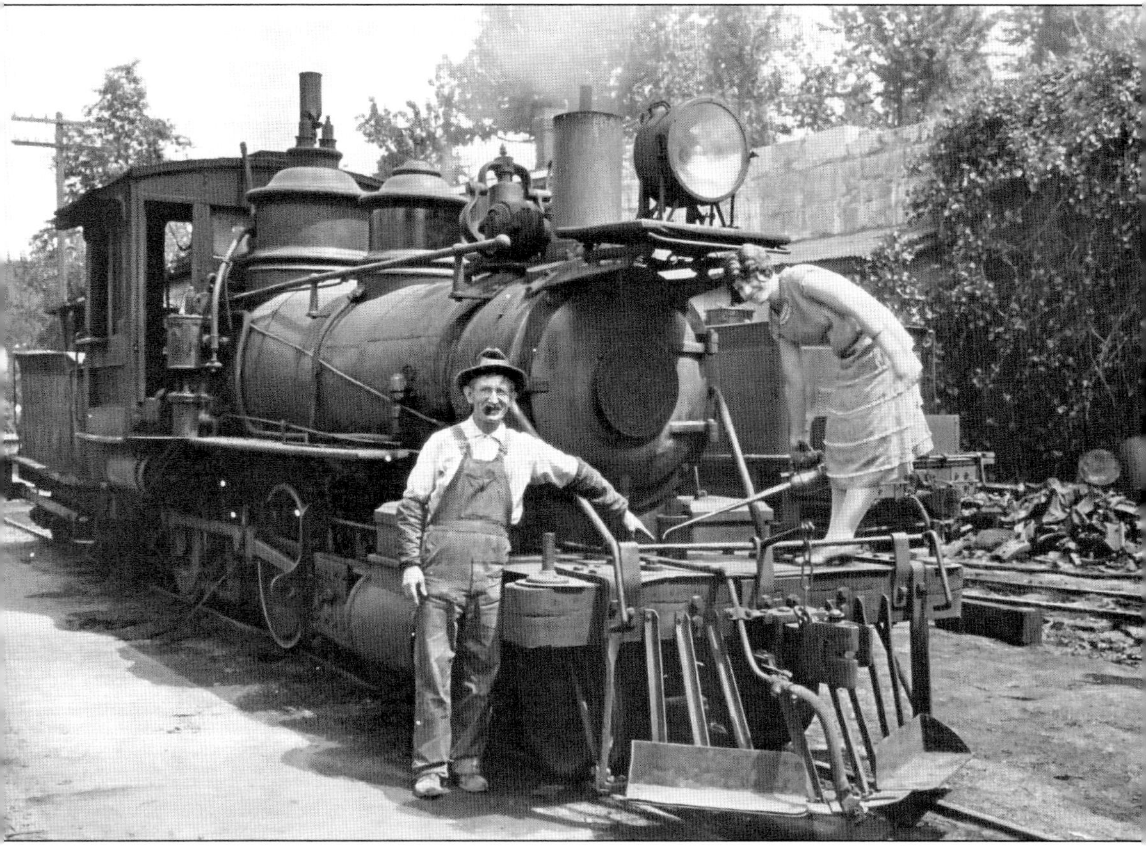

By 1906, once-plentiful timber had become precious in the area. Quantities of trees in the foothills had been downed for lumber to support the construction of the railroad, the mines, and town buildings. As wood was scarce, it was no longer an economical fuel for the railroad. In 1901, the first scientific excavation occurred at La Brea. The resultant fuel became cost-efficient once tank cars replaced wooden barrels for transport. By 1908, the railroad had retrofitted all of its wood-burning locomotives to use the cheaper, more accessible fuel. Oil had become the new energy source. In this photograph, freight agent Tom Kennedy poses with his bookkeeper, Sadie Geronimi Angiolini, who is "oiling up" narrow gauge engine No. 5. (CC.)

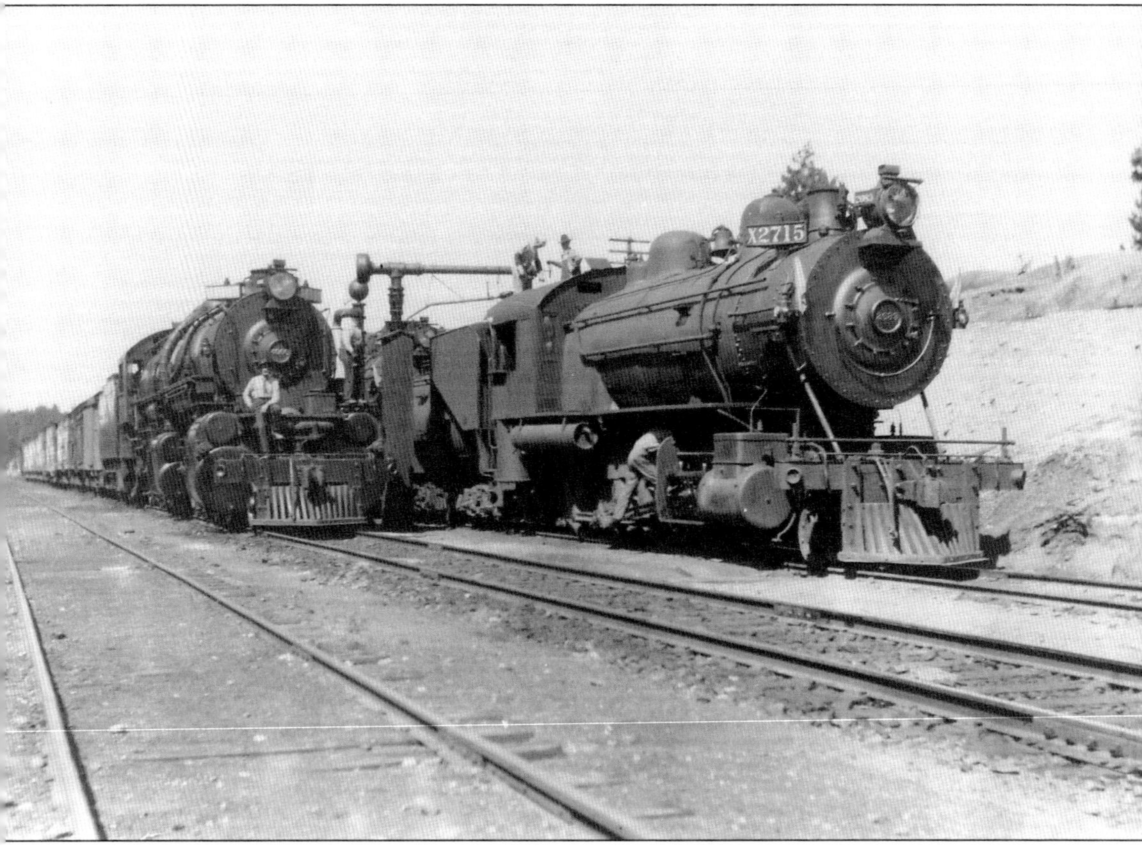

When carefully studying this photograph of the two steam engines and the three sets of parallel tracks, the difference in the width of the tracks is apparent. The Southern Pacific trains traveled on tracks with the rails farther apart. The standard gauge measured just a little less than five feet—56.6 inches—between the two tracks, as seen on the right. The tracks used for the Nevada County Narrow Gauge, on the left, had a space of only three feet, hence the name. (JLJ.)

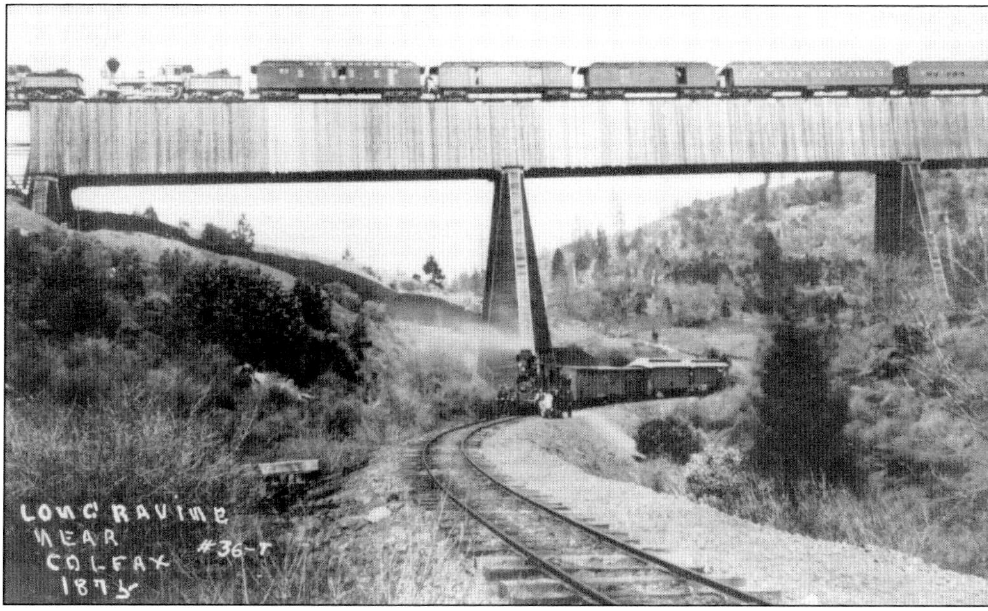

This photograph of the Long Ravine trestle was taken just beyond Colfax in the 1870s. Two long steel railroad trestles supported the train overhead. The original wooden bridge and trestle had been built in the 1860s. The NCNGRR trains traveled along the tracks underneath the bridge, while the Southern Pacific moved along the upper tracks. (AS.)

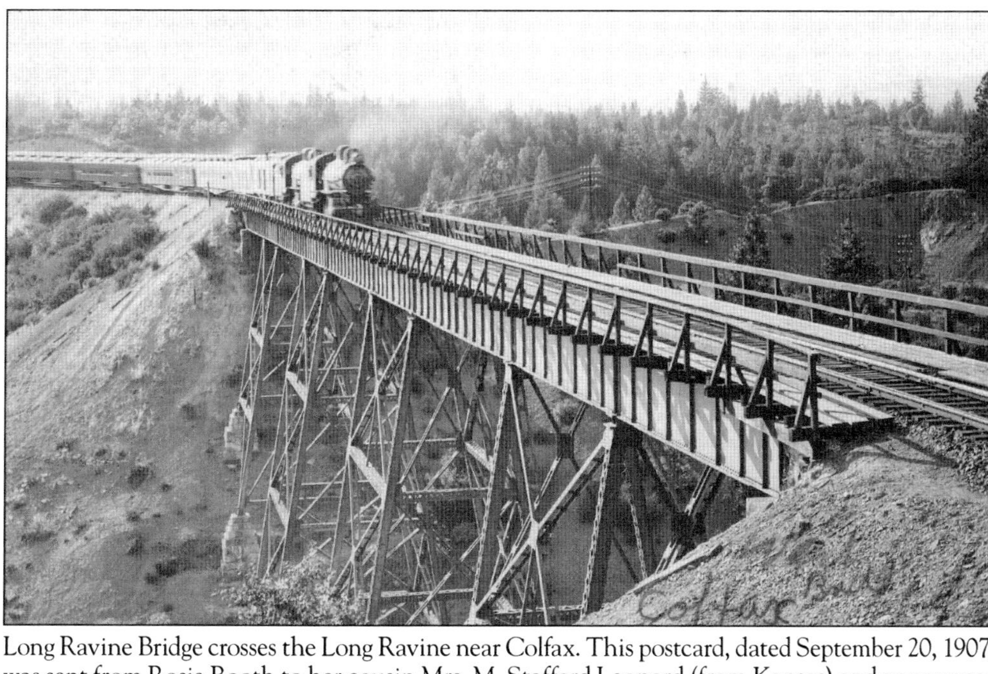

Long Ravine Bridge crosses the Long Ravine near Colfax. This postcard, dated September 20, 1907, was sent from Rosie Booth to her cousin Mrs. M. Stafford Leonard (from Kansas) and announces the birth of her baby boy, Clyde Newton-Booth. She asks her cousin to send her a "leather postal," meaning a postcard made of leather, as was customary at the time. (PCM.)

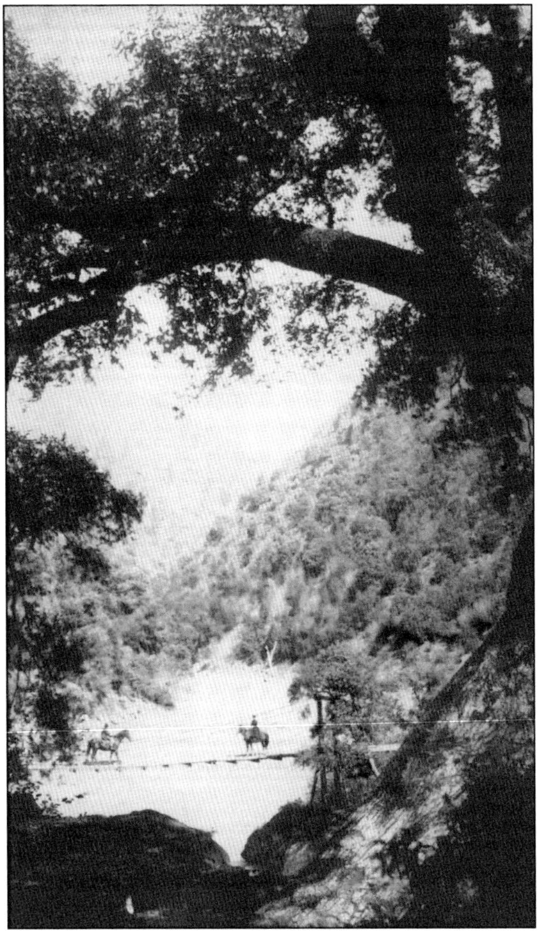

The view from the old bridge must have been exciting looking down the river canyon from atop a horse. Early on, people had few choices for transportation to their destination—walk, ride horseback, or take the stagecoach. When the railroad was completed, travel time greatly decreased while comfort increased. (CHM.)

In 1908, the new bridge over the Bear River stood 194 feet high, replacing the older wooden one, which, at 96 feet high, was considered the tallest railroad bridge in California. Made of steel, the new trestle stretched between the two tall banks, providing an incredible view of the river below. The Nevada County Narrow Gauge Railroad No. 9, made up of a combination of freight and passenger cars, crosses the bridge while passengers take in the full flow of the river underneath. (CHM.)

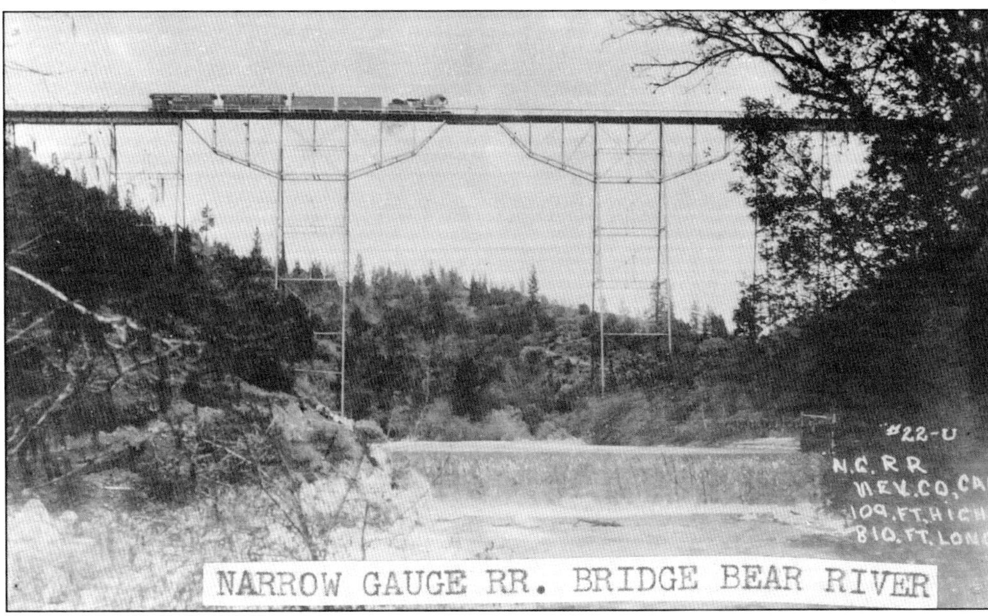

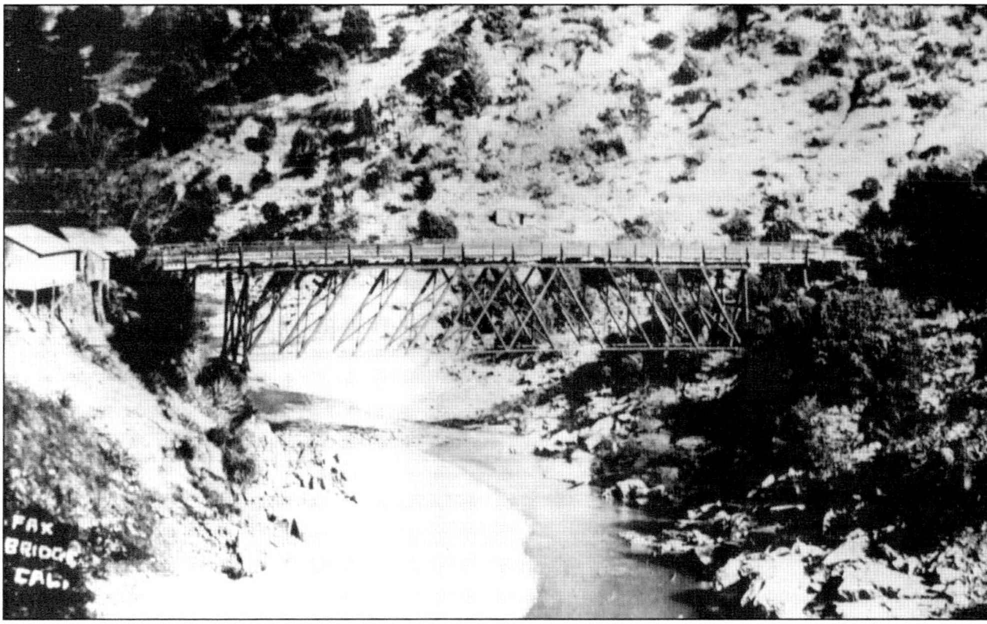

Charles Rice owned and operated the first ferry service at Mineral Bar. Later, he built a wooden toll bridge, enabling easier passage over the river. The photograph was shot from the Colfax side and shows the front of the tollhouse. Private parties built toll roads and bridges as businesses and charged every person and vehicle for the right to cross. (PCM.)

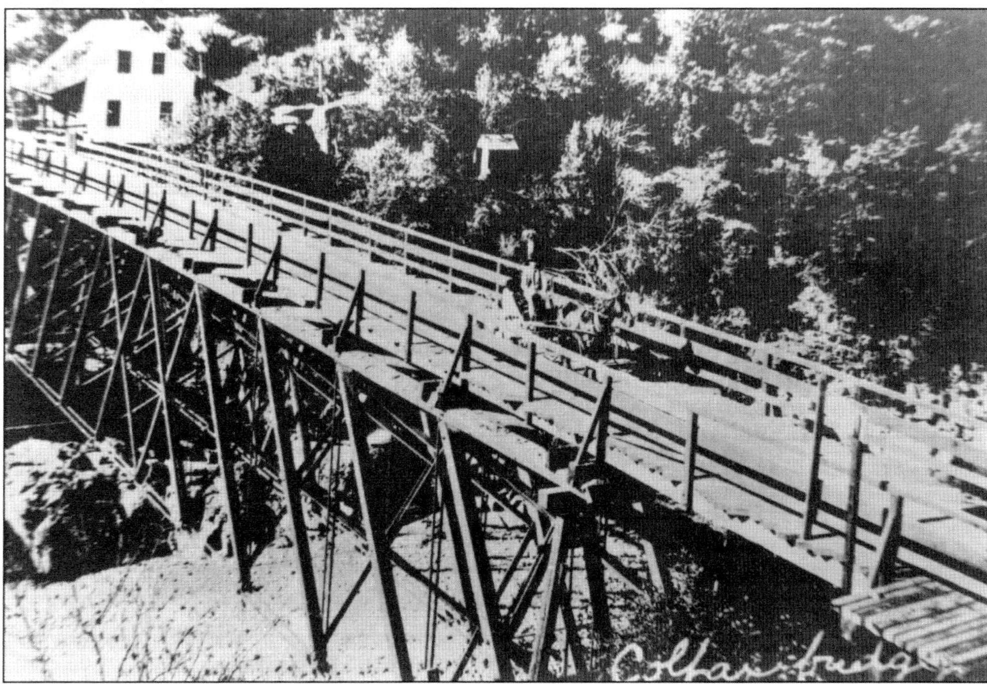

This postcard shows the Old Mineral Bar Bridge over the North Fork of the American River, which allowed travelers to cross the river en route from Colfax to Iowa Hill. The house in the background belonged to Frank Vennewitz Sr. and his son Frank Jr. They cared for riders' horses in their barn. In the early 1930s, the bridge collapsed, and it was never replaced. (PCM.)

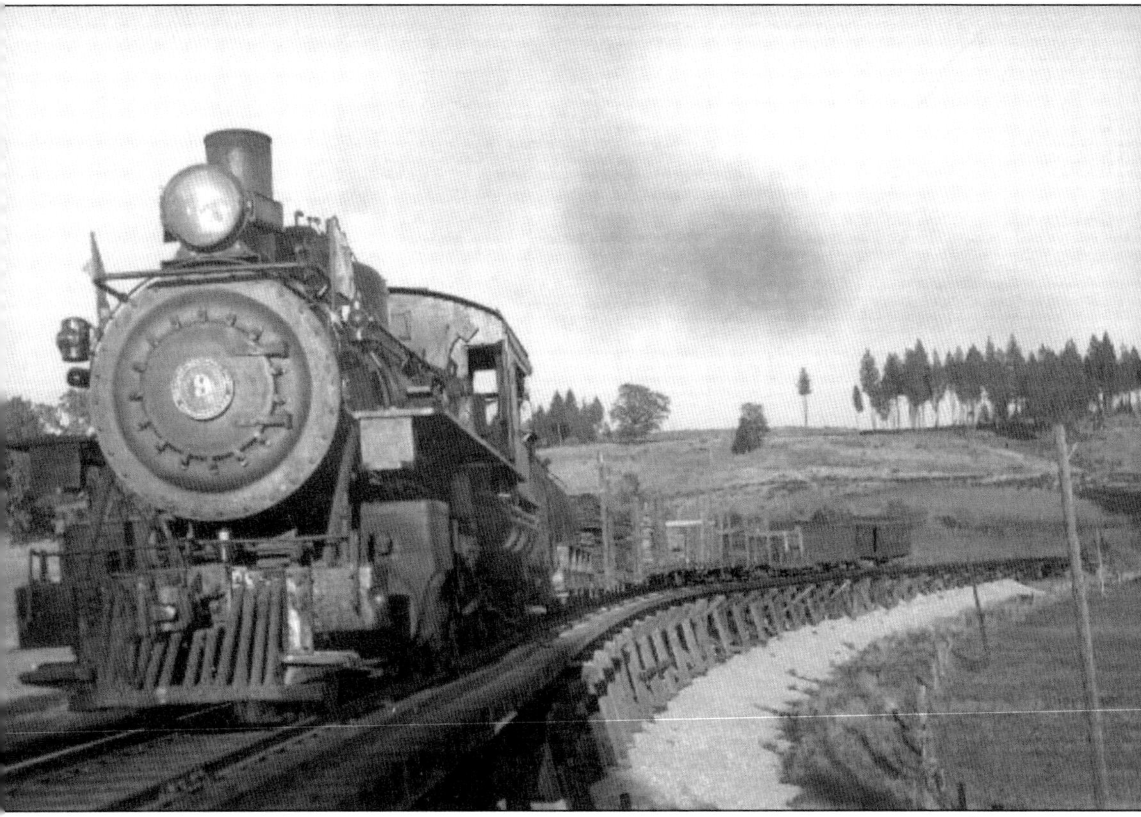

Below, the Nevada County Narrow Gauge Railroad train NG 2-8-0 No. 9 crosses the Jennings trestle, a long wooden trestle about four miles from Colfax, close to the Bear River Canyon. Fred C. Stoes took this photograph in 1937. (CHM.)

Four
EARLY DAYS

Colfax became a town after the Central Pacific Railroad established a camp there as a labor and materials distribution site for the construction of the transcontinental railroad. The company laid out the town and then sold the land to two Auburn merchants, Mr. Kohl and Mr. Kind. On July 28, 1865, the two gentlemen auctioned off individual lots to willing buyers. Merchants in Illinoistown wanted to take advantage of the new market and relocated their businesses to Colfax to be close to their customers. The new town took over as the main supplier of goods to miners and to remote locations.

Transportation consisted of stagecoaches, pack trains, and individuals on foot or horseback. Goods and mail reached the foothills and then got distributed in the same manner. People moving west seeking gold and land came primarily by coach or wagon until trains crossed the country.

The railroad changed everything about the way people lived and traveled. Population centers clumped astride railroad routes. Goods became available remarkably fast for the time, and heavy supplies for mining and construction became accessible. The western towns grew and stabilized, with commercial establishments offering much more than just the basics they previously sold. Distribution to individual customers not on the rail line continued via foot, wagon, horseback, or pack animal until the advent of the automobile.

It took the Industrial Revolution to bring the technology of electricity and modern communication to small towns, let alone to the average citizen. Wood-burning stoves for heating and cooking, along with outhouses, remained well into the next century. Water continued to be delivered by canals and flumes to mines and to farms for irrigation. Wagons brought water to businesses and to townspeople who did not have wells. However, communities like Chicago Park, Colfax, and Dutch Flat could ship their fruits, lumber, and other products by rail to the rest of the nation, thereby diversifying the local economy.

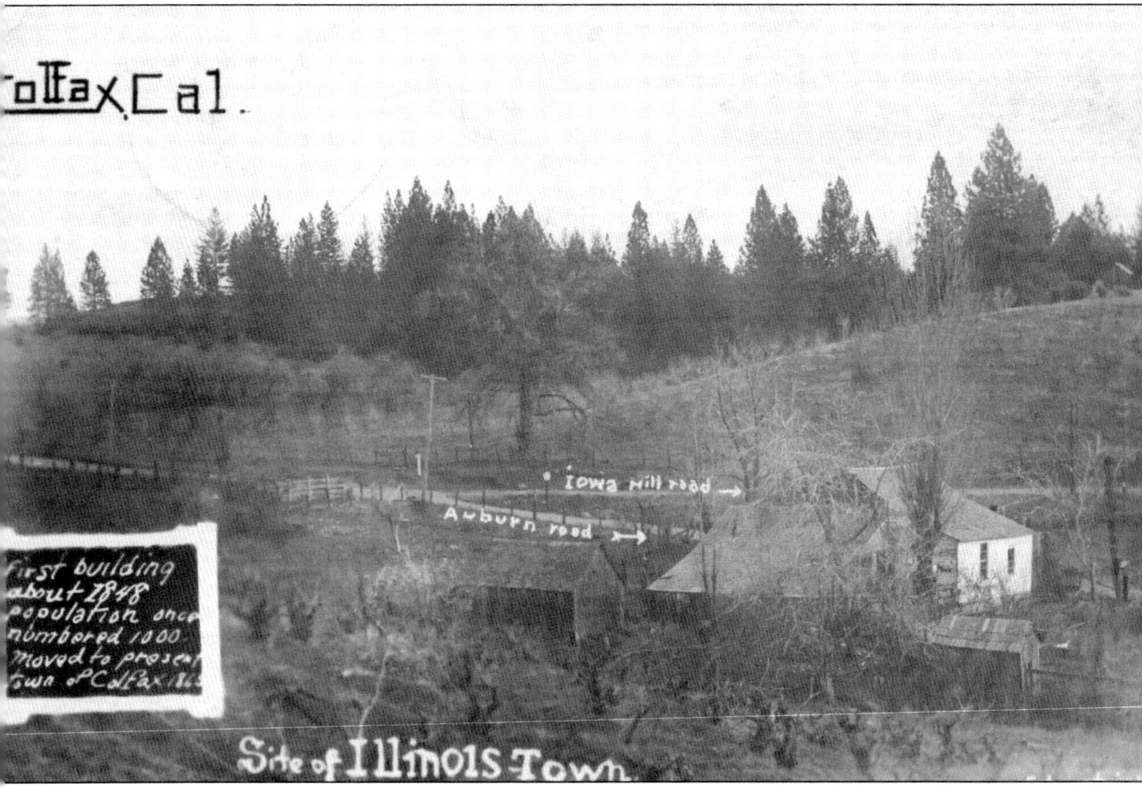

The front of this postcard shows Illinoistown with its first building, constructed about 1848, when the population totaled 1,000. The town was about 18 miles north of Auburn and a half-mile from where Colfax would upstage it upon completion of the railroad. The original town prospered from 1849 to 1865, supplying miners with staples and tools. Originally called Alder Grove because of its trees, the name changed to Illinoistown in October 1849, when a small group of Illinoisians elected to change it while downing a bottle of whiskey. A seven-mile turnpike built between there and Iowa Hill, referred to as Mineral Bar Bridge and Road, became essential to travelers, especially since Illinoistown lay on the direct stagecoach route from Auburn to Iowa Hill. This card was postmarked October 1914. (AS.)

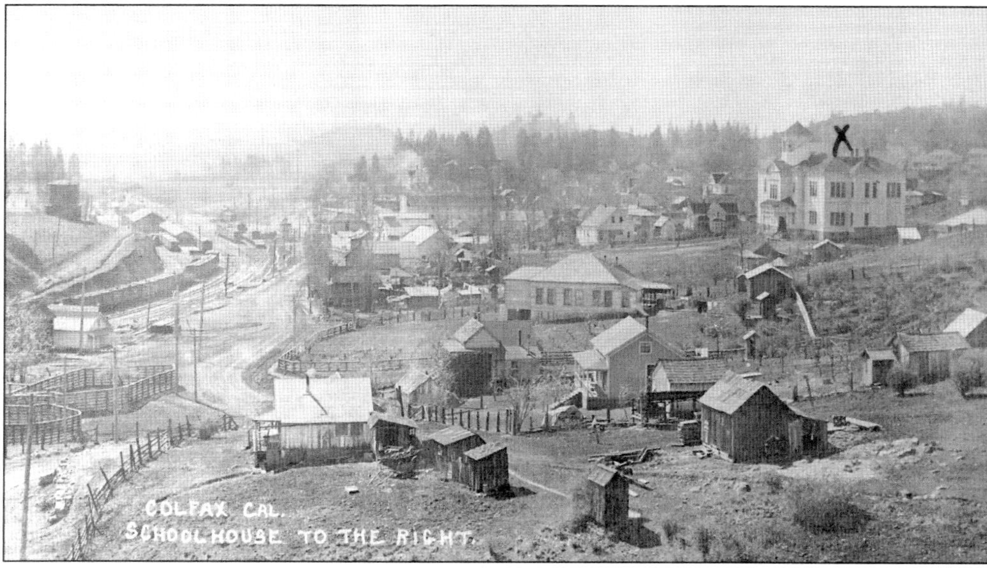

This early photograph shows a survey of Colfax and the water tower that serviced the railroad. The tower sits on the hill to the left of the train tracks, and the two-story schoolhouse (X) sits on the hill to the right. The engineers had the old tower removed to accommodate the 1910–1914 realignment of the railroad tracks to straighten the "Lazy S" configuration. (JLJ.)

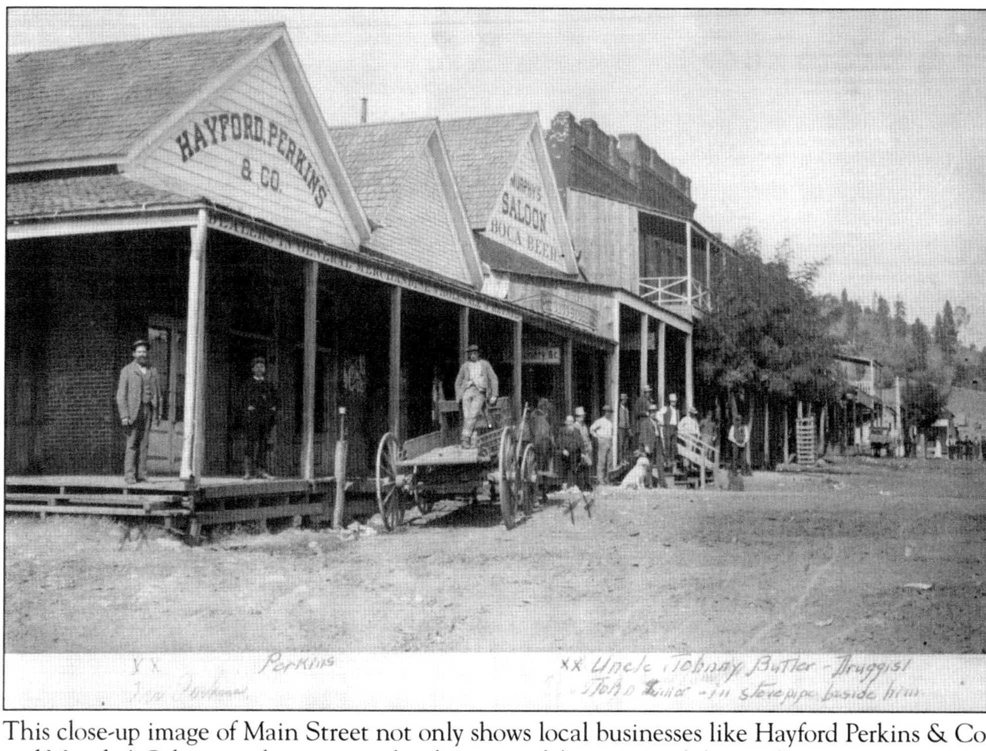

This close-up image of Main Street not only shows local businesses like Hayford Perkins & Co. and Murphy's Saloon, it also conveys the character of the town and the residents who lived there. At the end of the wooden sidewalk on the left stands Mr. Perkins. "Uncle" Johnny Butler is next to the wagon with the tall, spoked rear wheels and smaller front wheels. The shorter man in the dark suit was a druggist, and John Miller stands beside him with his stovepipe hat. (AS.)

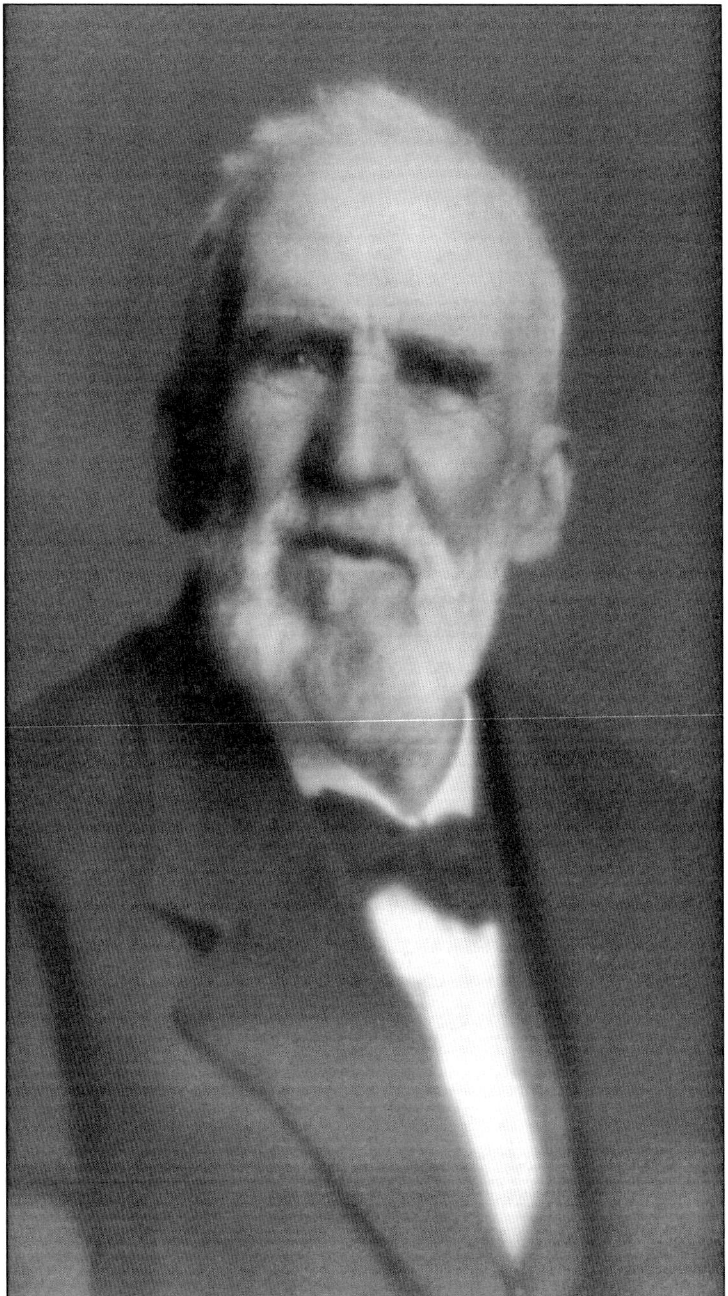

Ben Taylor, one of the earliest pioneers, arrived from Kentucky in 1849 and settled in what became Nevada County. Taylor lived at the Buena Vista Ranch in the Peardale area with a partner, Sandon Glassock. The two of them operated a store there. Taylor teamed up with brothers John and Edward Coleman, Gerry Morgan, and J.H. Neff to build a road for pack trains to connect Grass Valley to Illinoistown and the new railroad. The toll road and bridge, completed in 1865 at a cost of $60,000, was named Taylor's Crossing Bridge. It washed out in a flood and never got rebuilt, but Ben Taylor Road and Taylor Crossing Road still exist today. Credited with establishing the first toll road over the Bear River, Taylor provided convenient travel between the two areas. (SK.)

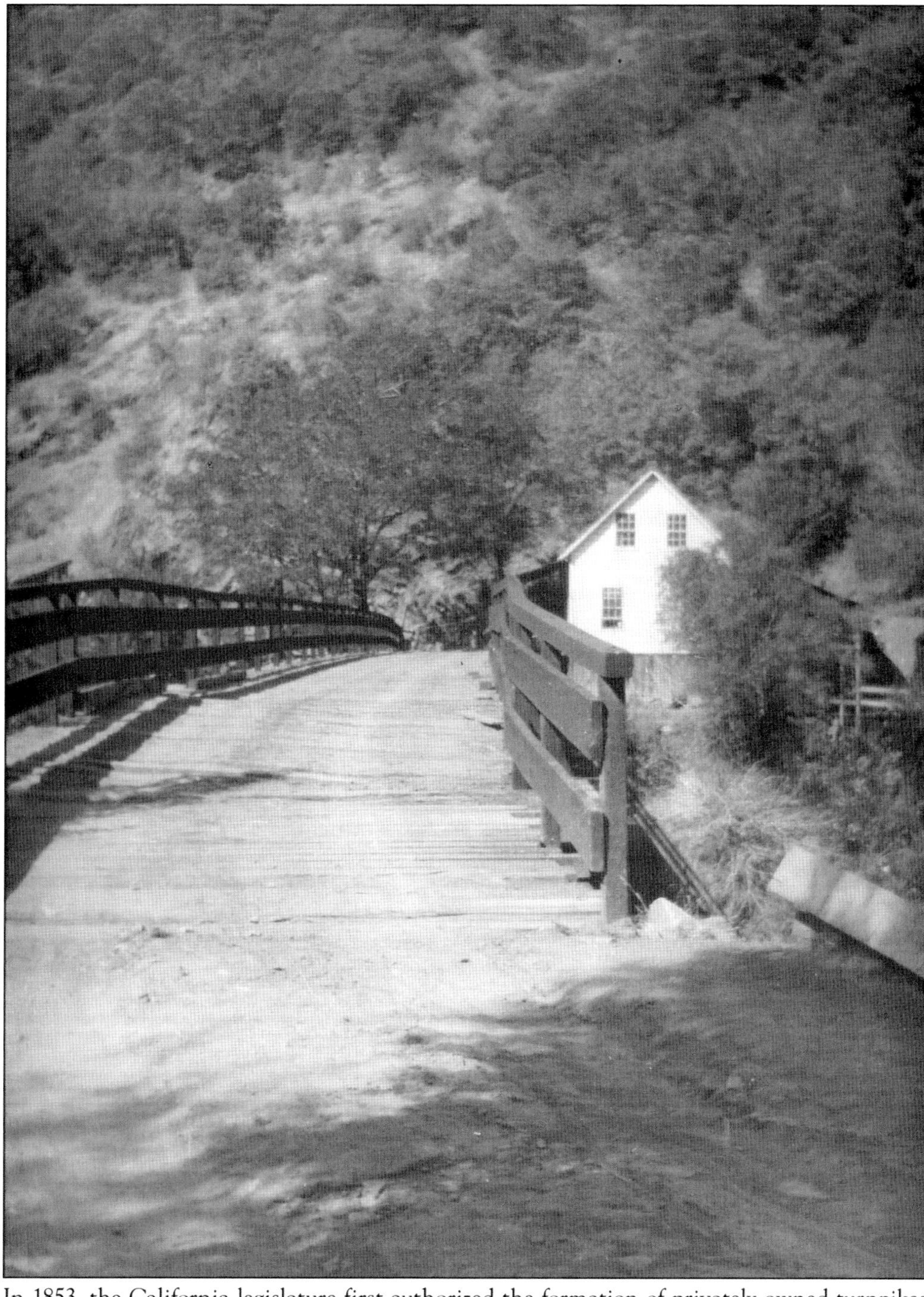

In 1853, the California legislature first authorized the formation of privately owned turnpikes as businesses. Also known as toll roads, the Sierra Nevada foothills had several that were built before the railroad gave a faster, more convenient route. The roads were privately held, and those who desired to get to the other side had to pay a fee to the owner. Y.J. Henry Henning ran his toll station out of the white building at the end of the bridge on the North Fork of the American River. (PCM.)

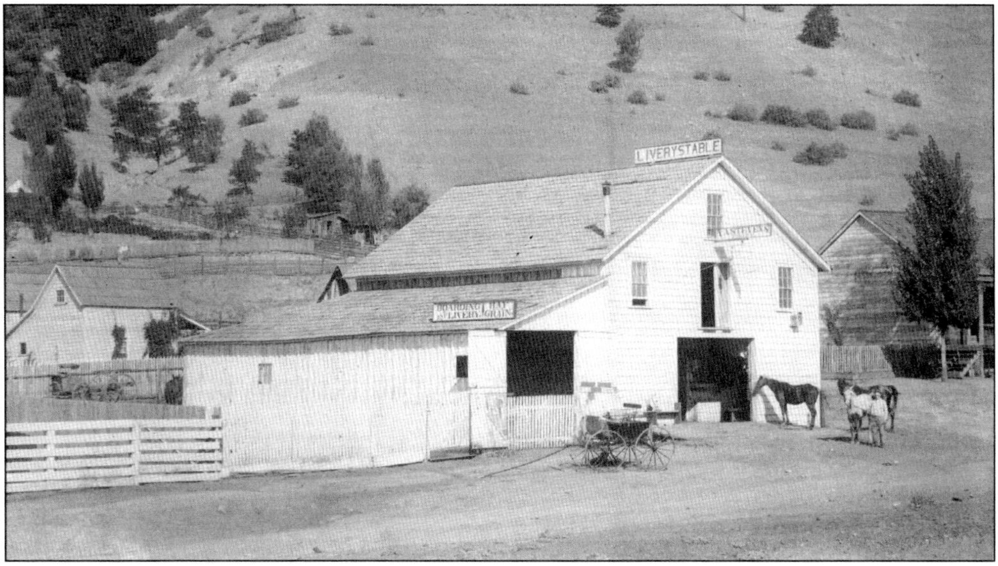

In the Wild West days, the role of the livery was the care and feeding of the horses boarded there during the rider's stay. Truman Allen Stevens owned one of the liveries in Colfax and provided "boarding, hay, livery and grain." He operated the business for eight years before his son, Amos Stevens, who was a judge, took over the livery. The older Stevens owned a ranch in Iowa Hill and delivered mail to gold camps. W. John Rutherford built Stevens Trail, which began as a toll path from Colfax to Secret Ravine and then on up to Iowa Hill. It was used from 1870 to 1895. (CHM.)

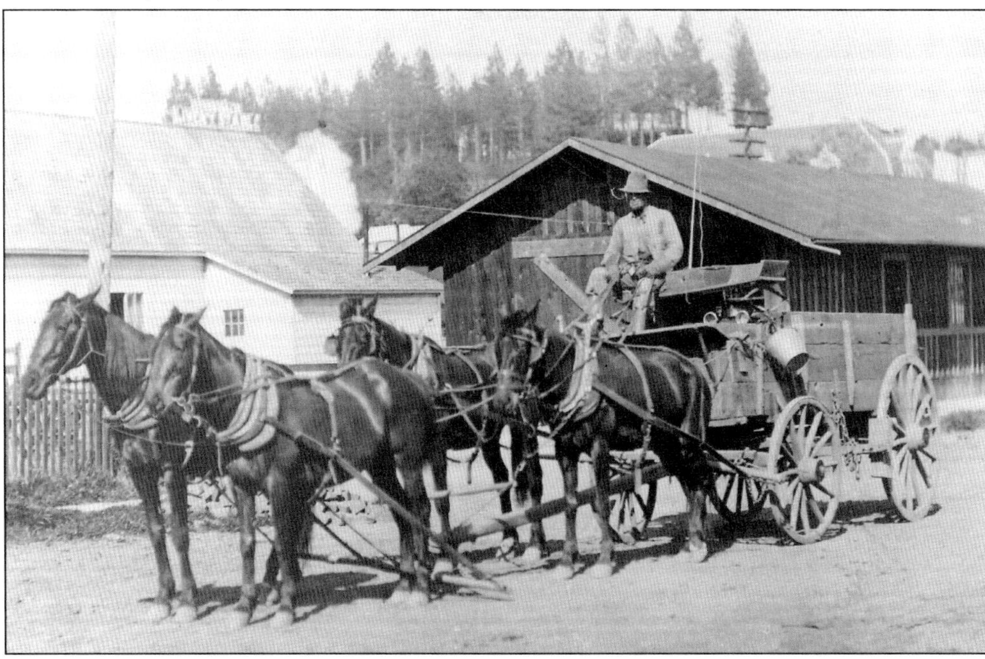

Russell's Livery operated within the town of Colfax. This photograph shows the driver perched on the wagon, drawn by his team of four mules. He looks ready to set off for a delivery or a pickup. The World War I Memorial Hall was later erected on the site, and Colfax City Hall is now here as well. (JLJ.)

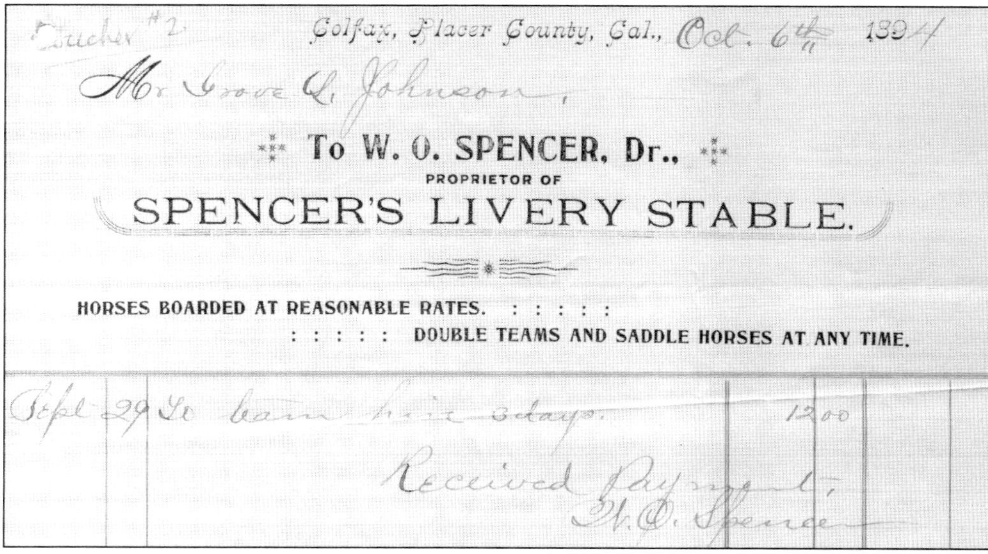

Voucher #2

Colfax, Placer County, Cal., Oct. 6th 1894

Mr. Grove L. Johnson,

To W. O. SPENCER, Dr.,

PROPRIETOR OF

SPENCER'S LIVERY STABLE.

HORSES BOARDED AT REASONABLE RATES. : : : : :

: : : : : DOUBLE TEAMS AND SADDLE HORSES AT ANY TIME.

Sept	29	To barn hire 3 days.	12	00

Received Payment,
W. O. Spencer

W.O. Spencer owned another livery in Colfax, Spencer's Livery Stable. According to this receipt, Spencer charged $4 per day in 1894 for the boarding of a Mr. Johnson's mount(s). The bill does not indicate the nature or number of the boarded animals. When the stagecoaches came through town, riders likely secured fresh mounts from one of the liveries in town. (AS.)

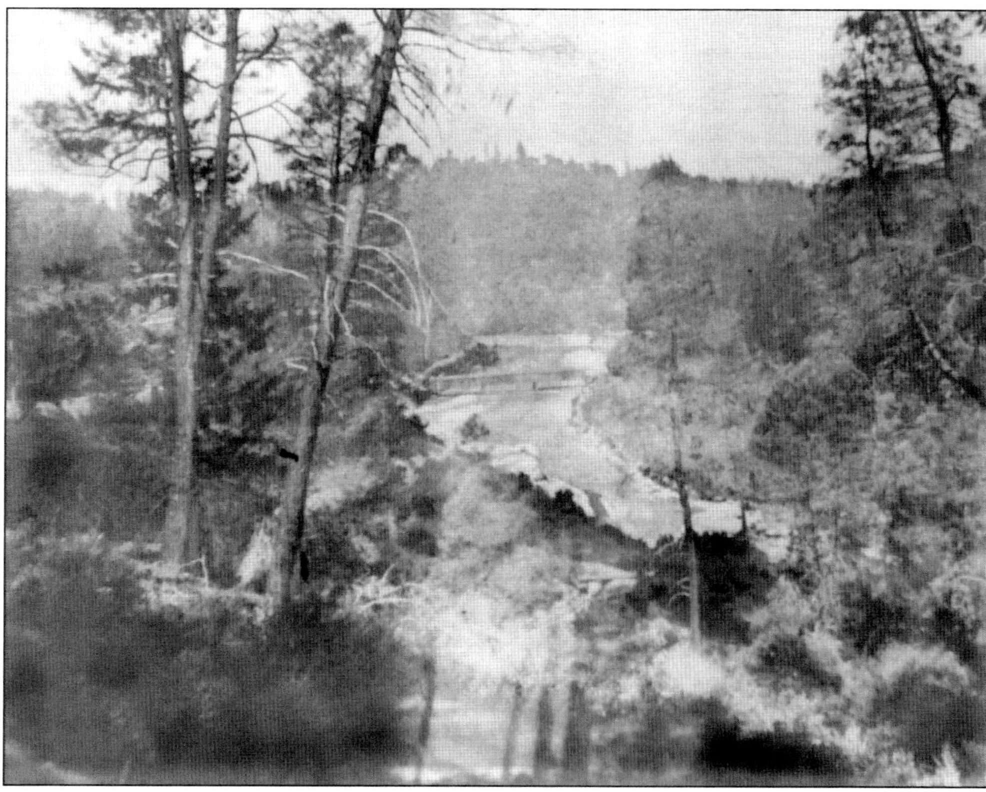

This 1866 photograph titled *Stage Road from Colfax to Grass Valley* faintly shows the old Bear River Bridge (center), which served as the crossing for stagecoaches, pack teams, and wagons. The bridge connected the road from Colfax to Grass Valley on the other side. (LOC.)

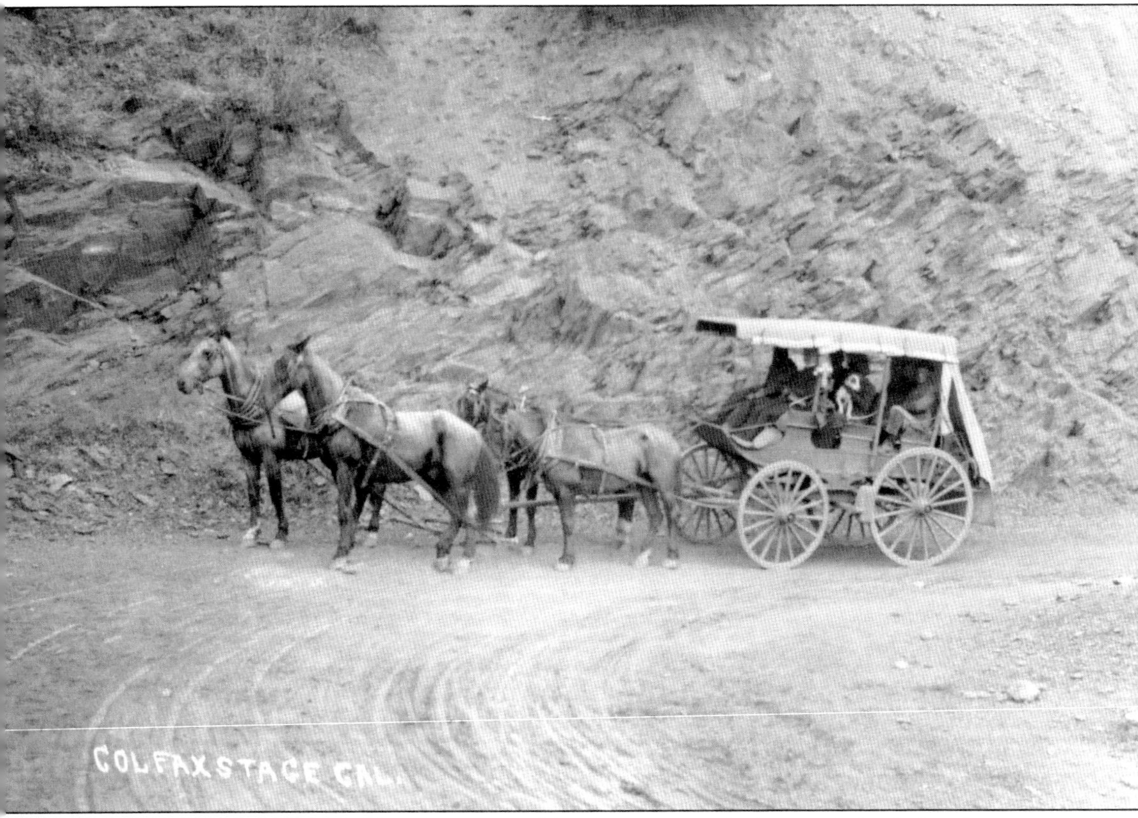

The Colfax Stage operated with a team of four horses to pull the coach loaded with passengers up the steep, windy dirt road between Colfax and Iowa Hill. The straight distance between the two towns is just over five miles, but that was not possible. Today, the road runs almost 54 miles between the two points and takes just under two hours by car. On this particular day, in about 1905, four passengers and a dog made the trip. Rides were both bumpy and dusty, but travelers did not have many options in those days. The stage was restored by the Native Sons of the Golden West as a treasured part of Old West history. The Bernhard Museum in Auburn has it for display. (PCM.)

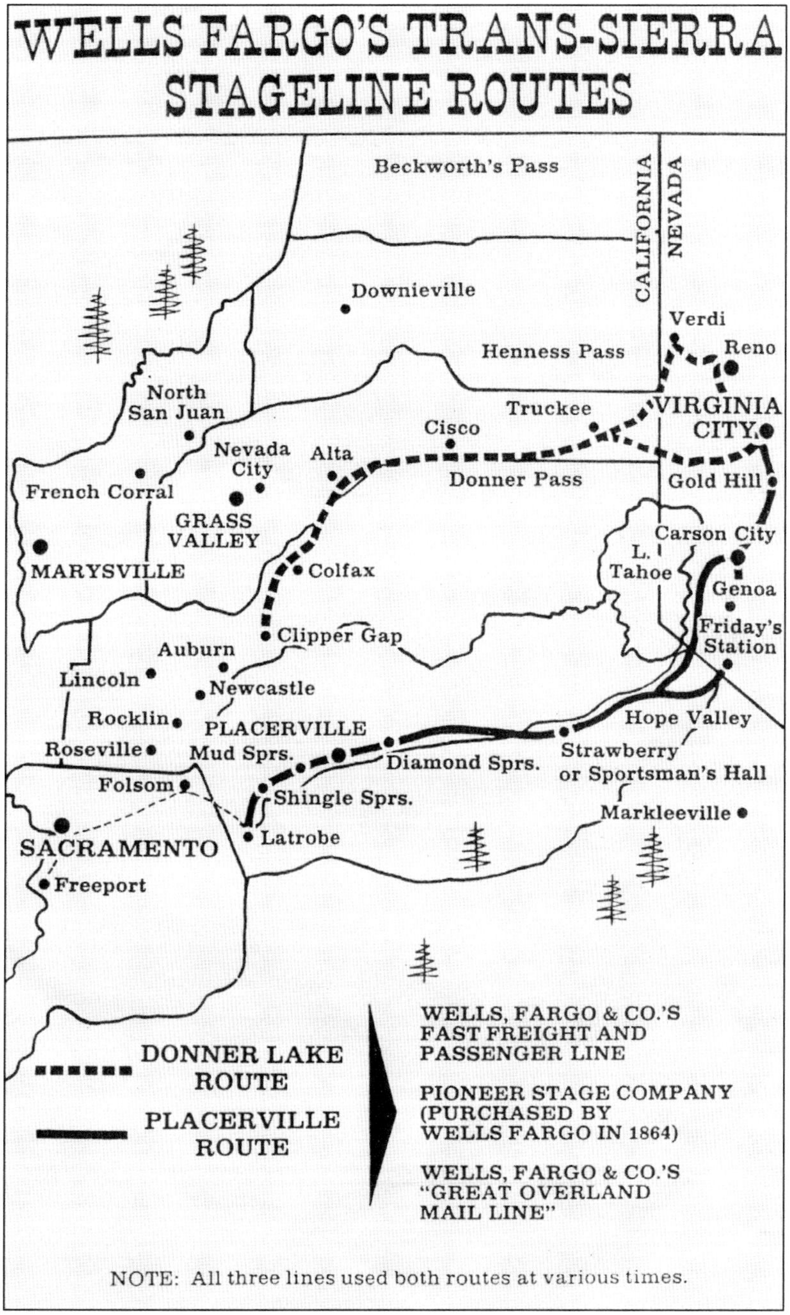

The map outlines the stage-line routes these three companies used to cross over the Sierra Nevadas. There were two routes. The one going through Colfax, about 13 miles from Clipper Gap, stopped at various small towns and then looped around via Reno and Virginia City. The southern route traveled from Virginia City southwest to Lake Tahoe and ended up a little east of Sacramento. Wells Fargo owned all three trans-Sierra companies listed here. The stagecoaches offered a vital form of transportation in the early days of the west. "Grandma" Nettie Keck was the first woman to arrive in Colfax by stagecoach. She was the wife of Benjamin Keck, who ran a harness shop in Colfax. (CHM.)

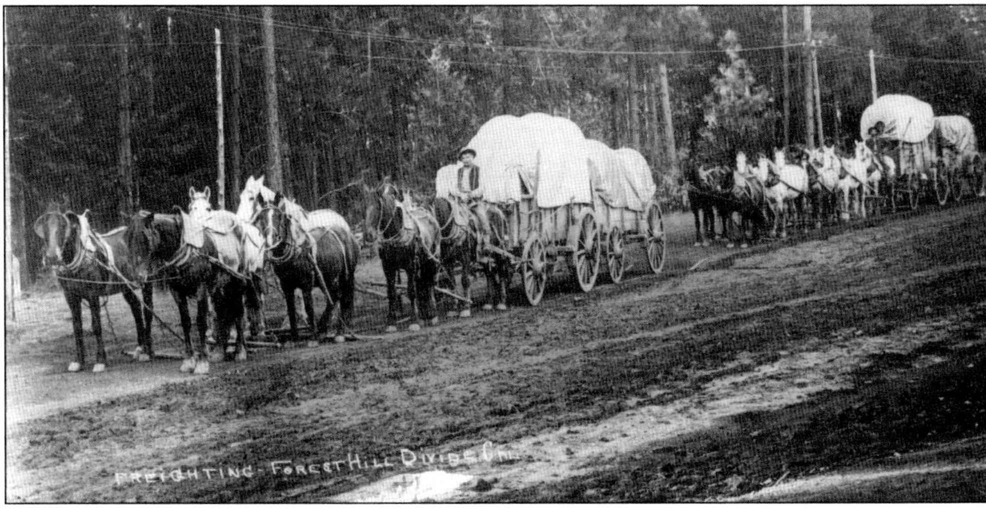

Large horse-drawn wagons, hitched to smaller wagons, freighted goods on the Foresthill Divide. Both pack trains and stagecoaches used some of the dirt trails and roads along the divide between the North and Middle Forks of the American River. (JLJ.)

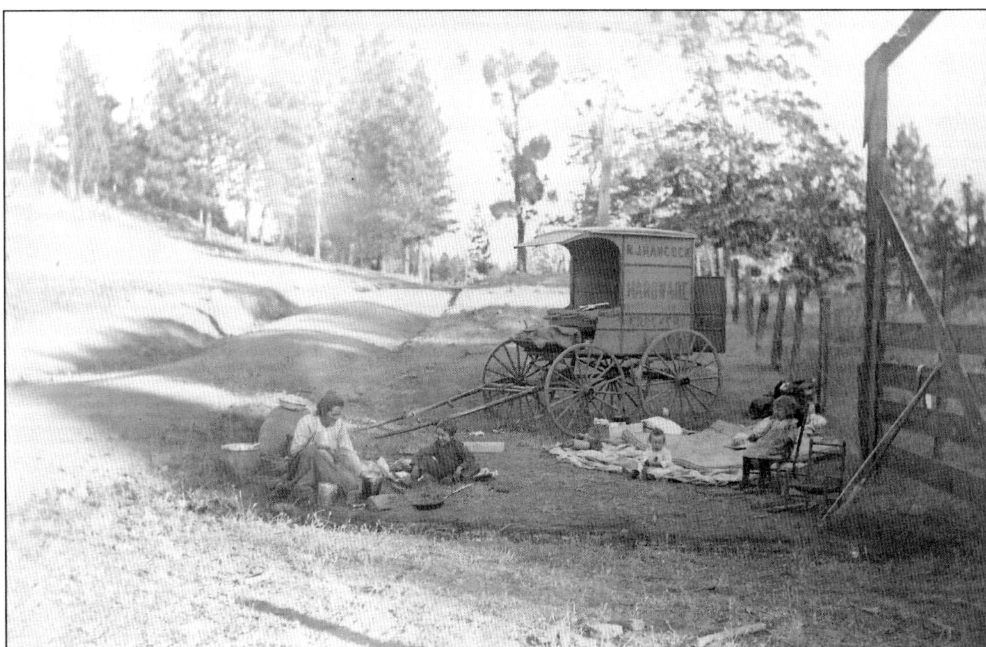

The R.J. Hancock hardware wagon delivered wares to its remote customers via this petite store on wheels. Traveling sales wagons offered direct access to goods. (JLJ.)

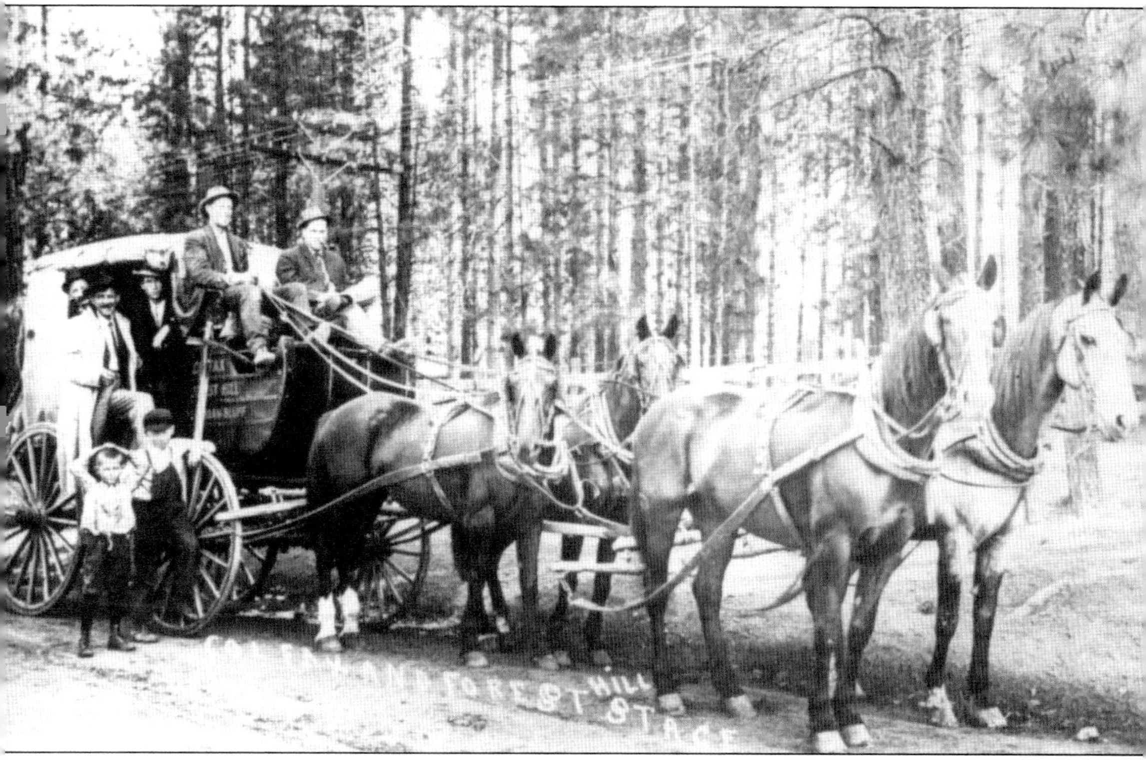

This close-up of the Colfax and Forest Hill Stage shows a full load of passengers pausing for a photograph at Iowa Hill. Two young boys lean up against the wheel of the coach, posing with the rest of the travelers. The coaches were built to withstand the rough terrain. The Concord wagon, the preferred design, used hefty leather straps that shaped the braces attached to the front and back axles. This structure eliminated most of the jerking and allowed the coach to rock back and forth, which was more comfortable. The stage lines had different owners depending on the route. The Colfax and Grass Valley Telegraph stage made two trips daily to connect with the railroad, priced at $2.50 for Grass Valley and $9 for Forest City. (CHM.)

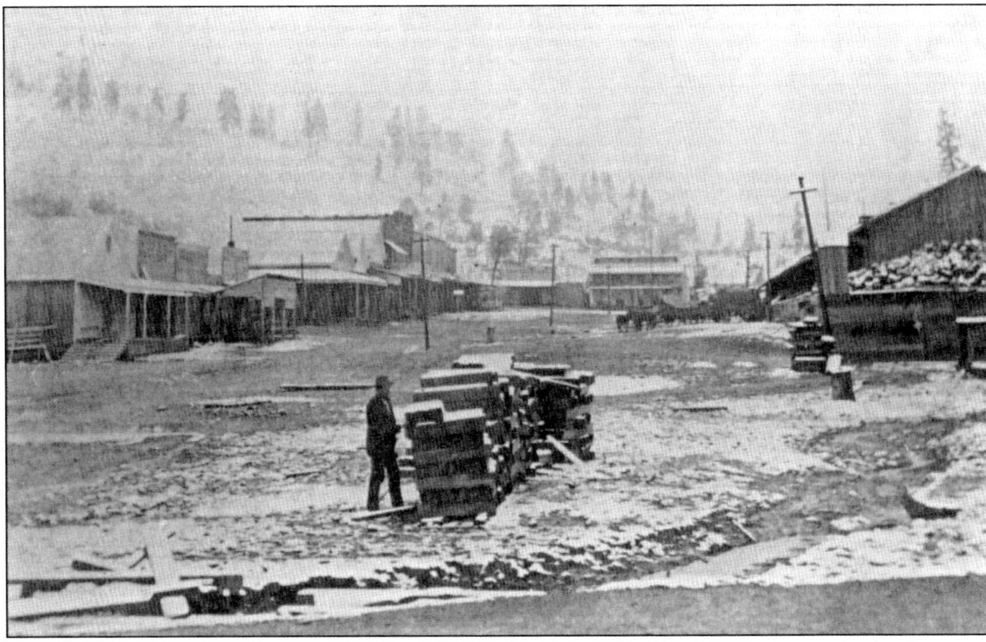

This wintry scene of Main Street in 1868 shows the various Colfax businesses in the background as the snow continues to fall and the trees on the hillside become almost invisible. (CHM.)

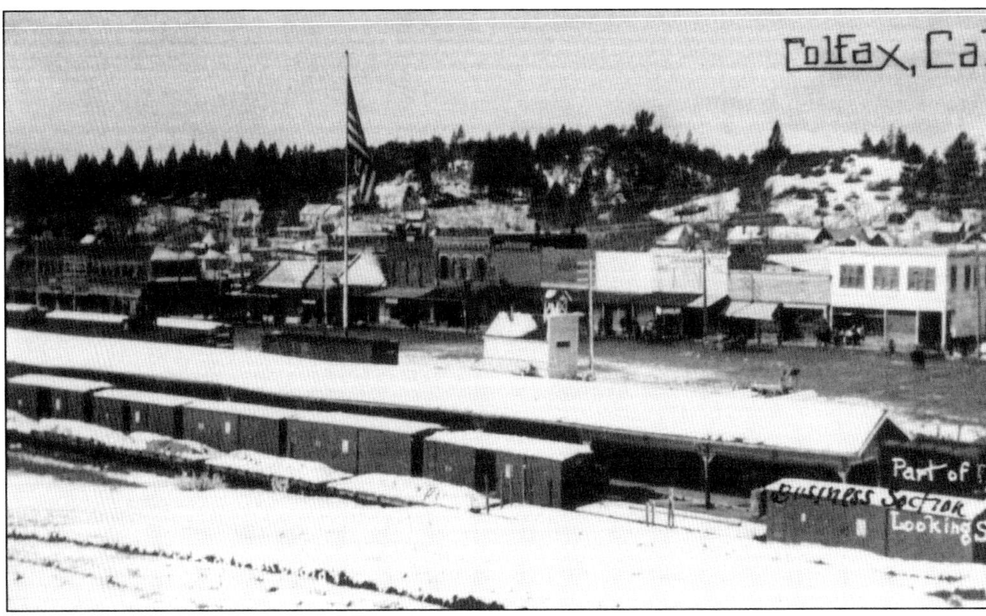

This 1937 scene shows snow heaped on the trains, tracks, and buildings. At higher elevations on the pass, sheds covered the tracks to protect trains from heavy snowfall. In May H. Southern's January 29 journal entry, "The Hard Winter of 1889–1890," in *The Covered Wagon*, she writes, "Few trains blockaded between Blue Canyon and Shady Run; 1,800 men with picks and shovels cutting away snow banks twelve to fifteen feet high; snow on sheds at Summit and Tunnel 13 from 50 to 250 [inches] deep and in isolated places 500 [inches]. Three hundred men employed extra between Sims and Edgewood with provisions scarce between these points. 3,500 men employed extra, of these 2,500 men on Central Pacific Railroad." (CHM.)

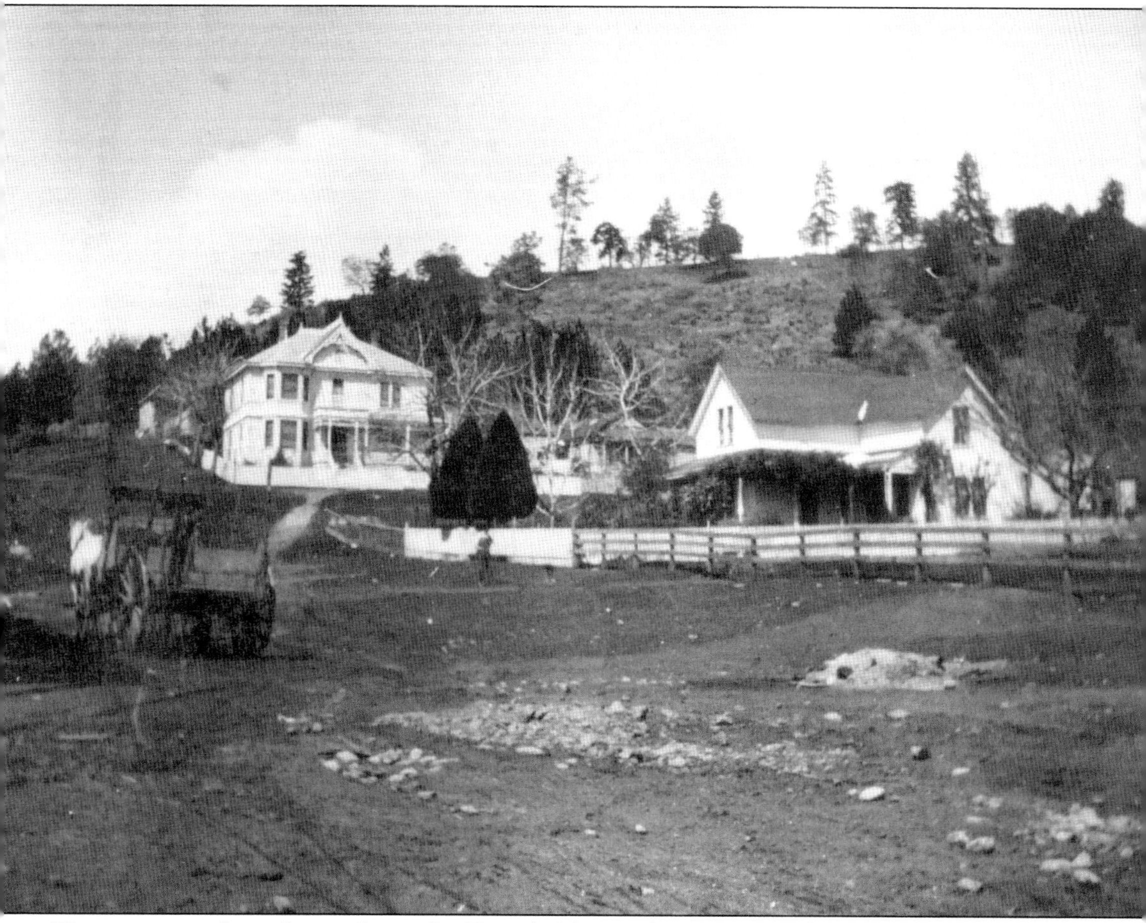

This sparse residential area shows the Kuenzlys' large white Victorian house at the corner of Depot and Pleasant Streets. John Jacob Kuenzly built the 4,000-square-foot home in 1897, and it still stands today. The Kuenzly family arrived in California during the Gold Rush and settled in Colfax in 1871. John Jacob Kuenzly had originally emigrated from Switzerland in 1854. He learned telegraphy and worked as an operator at Western Union for 15 years before switching to real estate. He became postmaster and justice of the peace. He also operated the general store beside the Corner Market. The home is often referred to as the Kuenzly-Lobner residence because Kuenzly married Maria, one of the Lobner daughters. They had a son, Frank. (AS.)

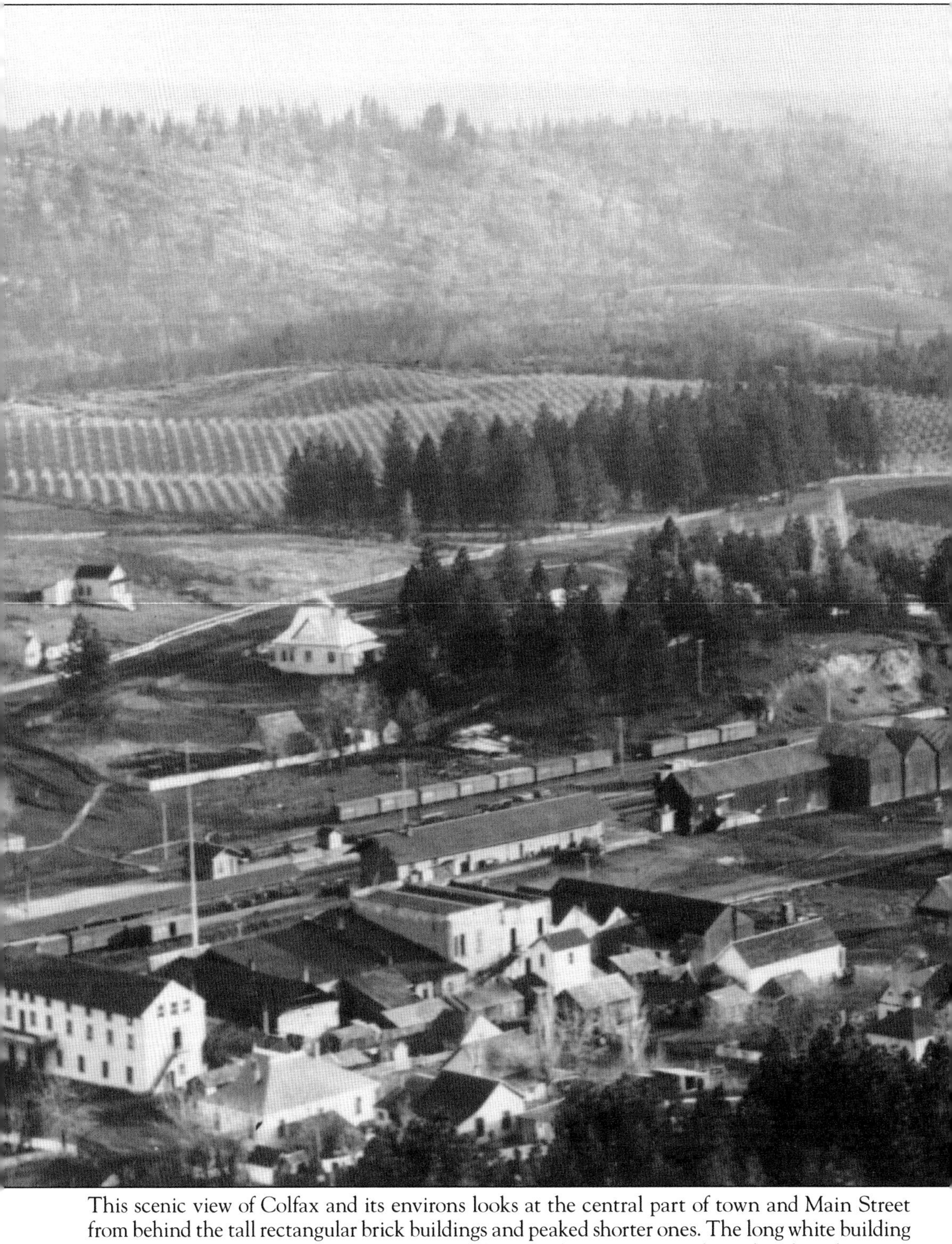

This scenic view of Colfax and its environs looks at the central part of town and Main Street from behind the tall rectangular brick buildings and peaked shorter ones. The long white building at the bottom left is the Mountain View Hotel. A freight train sits on the railroad tracks across from the passenger and freight stations near the center of the image. The Victorian home of the

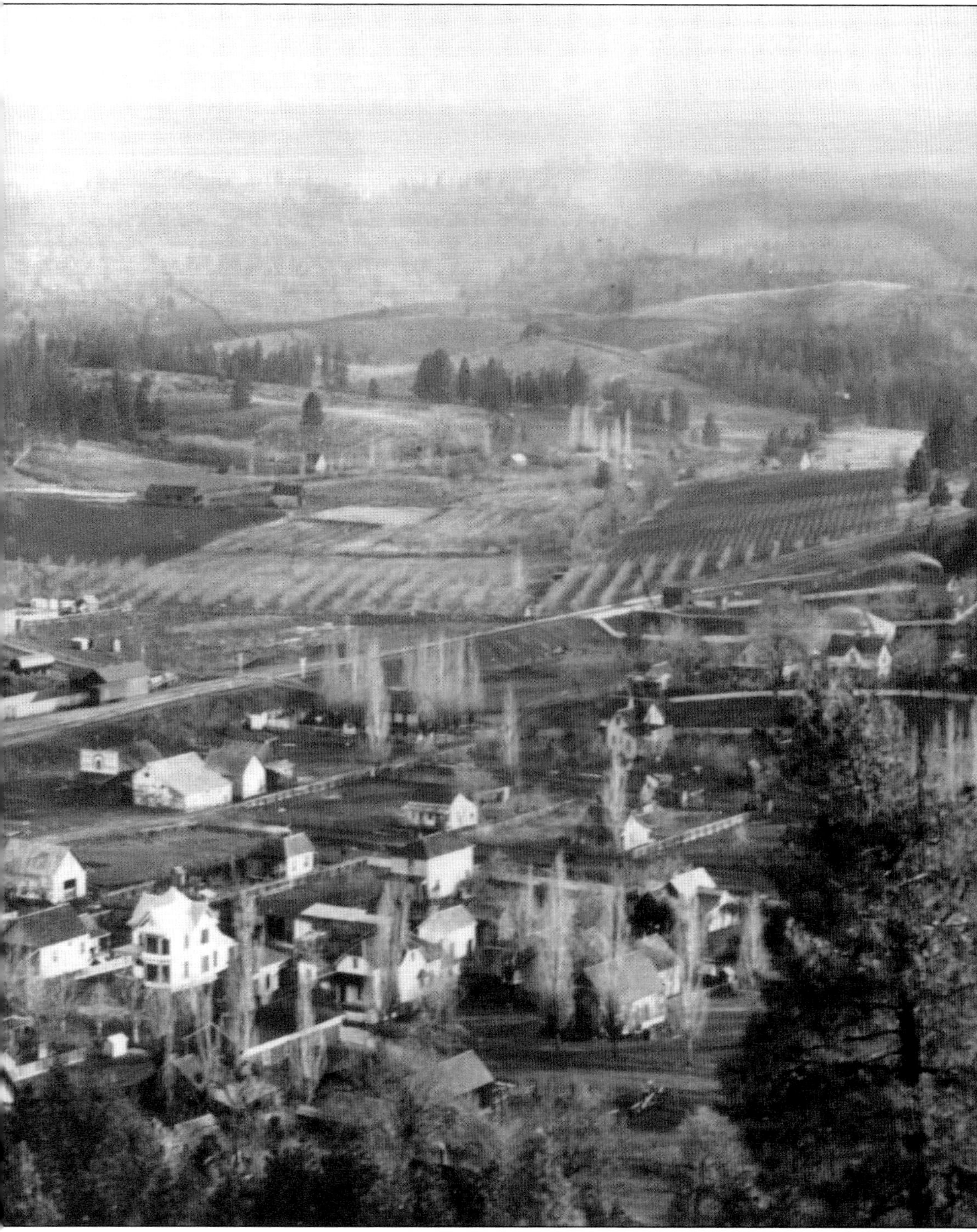

Kuenzlys stands up proudly at the bottom right. The expansive farms and hillsides beyond the town center reach far into the distance, with rows and rows of fruit trees and grape vines covering the hills surrounding the community. (CHM.)

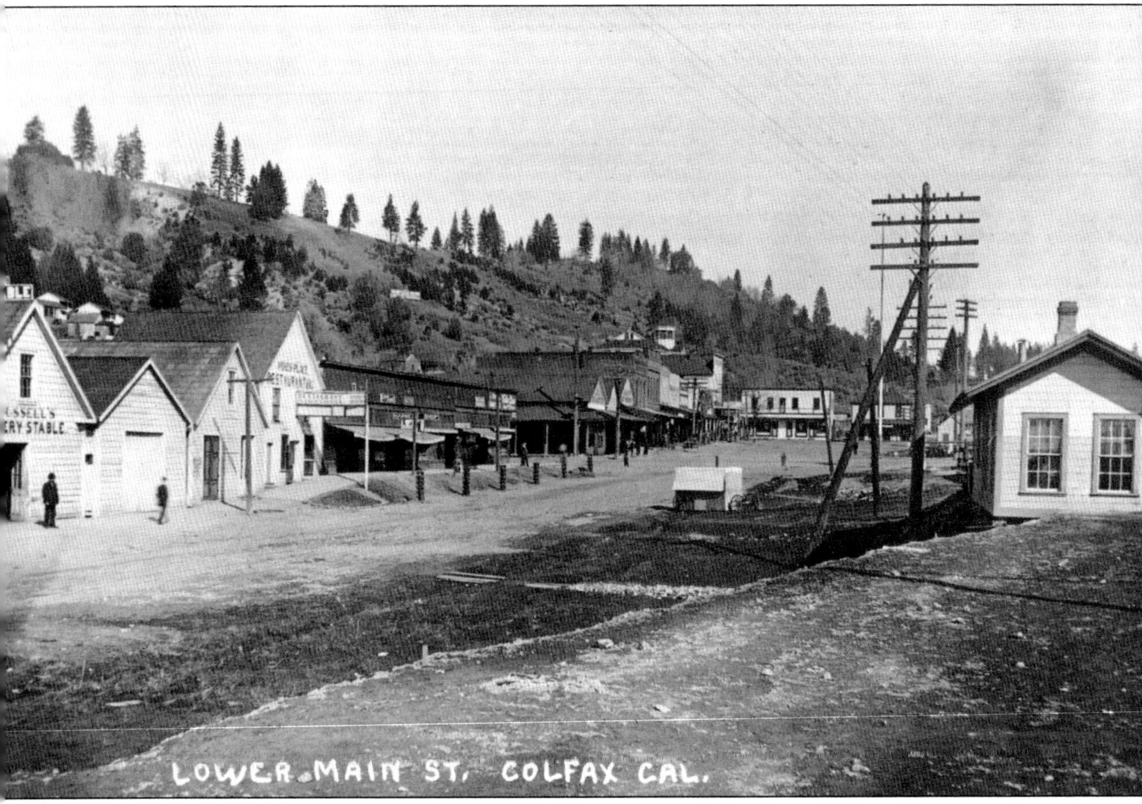

This close-up of downtown Colfax gives a true portrait of the foot of Main Street in the early days, with the peaked false-front buildings all in a row. The building at the lower left, Russell's Livery Stable, housed horses for locals and those just traveling through. Three buildings up the street, the restaurant and lunch counter has the one highest peaked roofs on that stretch. Next is a harness shop, with other merchants filling in the remaining short buildings on the block. The peaked structures resume again farther up the wide dirt roadway all the way to Depot Street at the end of the block. (AS.)

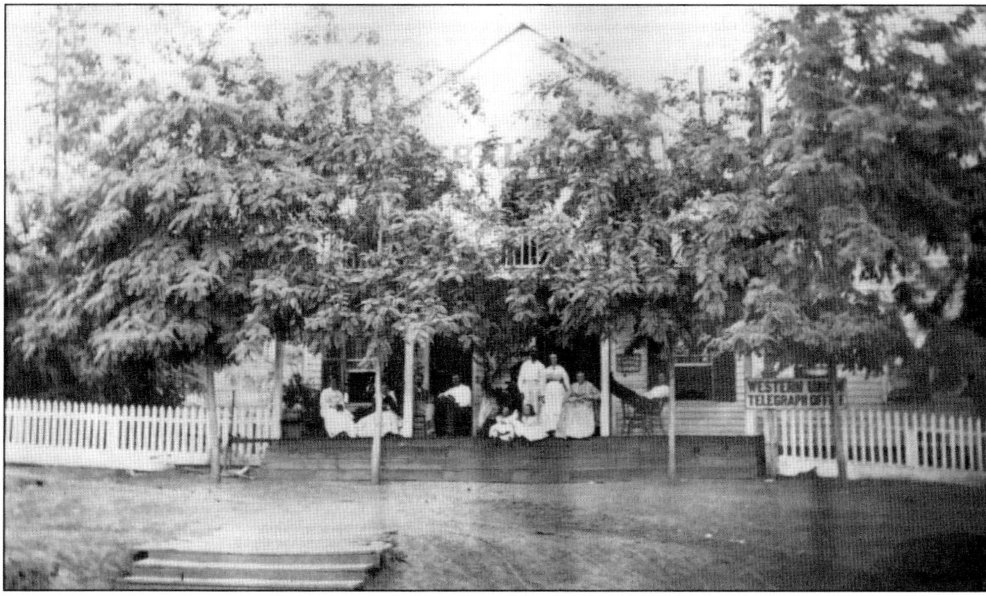

Peeking out from behind the dense foliage is the Central Hotel in Colfax. In this image, guests casually gather together on the porch to enjoy a pleasant afternoon. Thomas Curley and James Mahon were the proprietors of the hotel. The prominent Victorian Kuenzly house stood to the left of this hotel. (AS.)

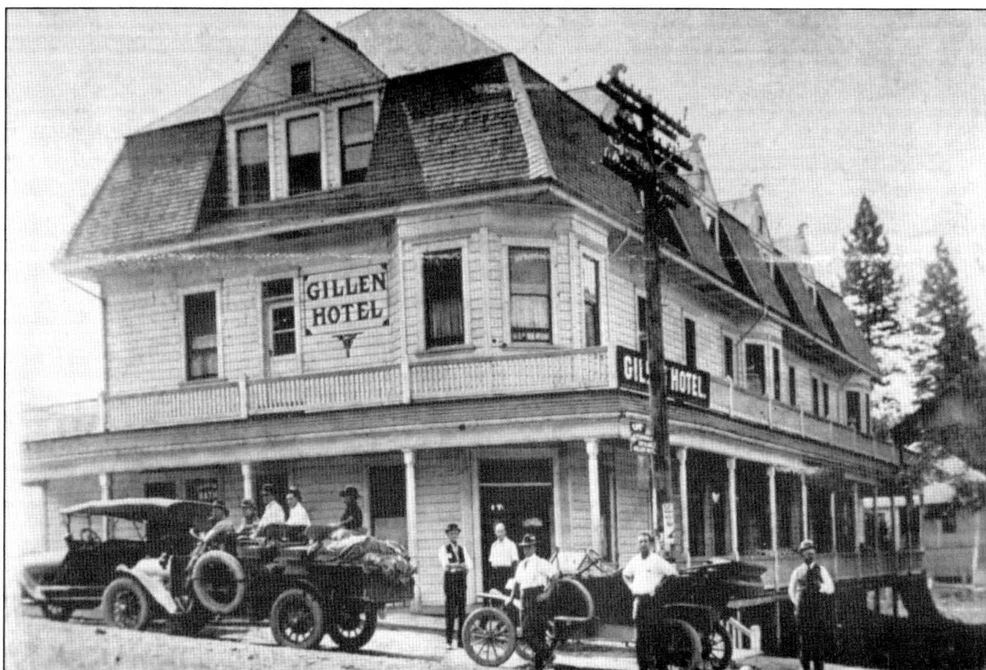

The Gillen Hotel, built by Daniel Gillen in the late 1890s, became the Colfax Hotel after his death. It cost him $25,000 to build and included an 80-foot deep well. R.F. Willis served as the contractor and architect. Gillen built a hall at Auburn and Grass Valley Streets where dances were hosted and town officials conducted meetings. He built his own power plant, the first in town, to light up his hotel and dance pavilion. (CHM.)

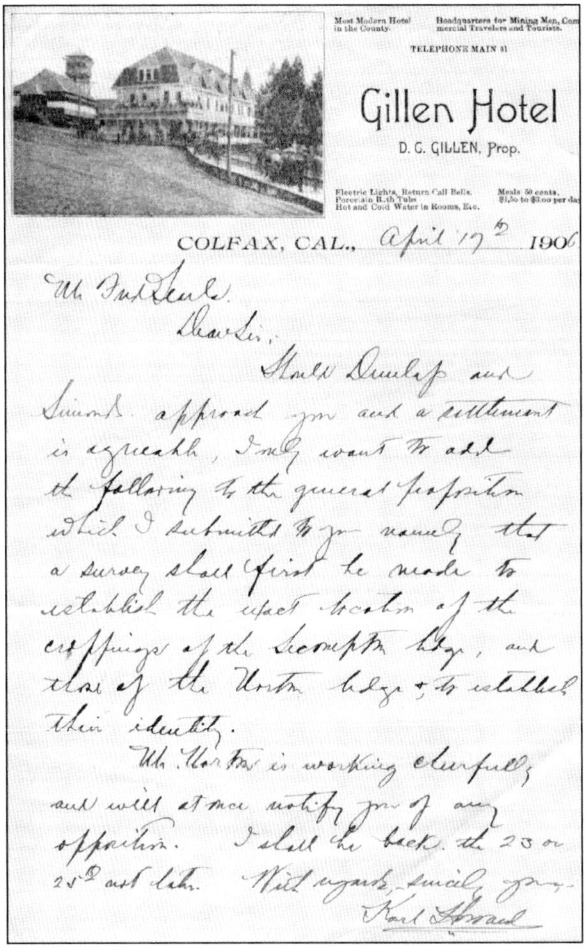

Most Modern Hotel in the County.

Headquarters for Mining Men, Commercial Travelers and Tourists.

TELEPHONE MAIN 81

Gillen Hotel

D. G. GILLEN, Prop.

Electric Lights, Return Call Bells, Porcelain Bath Tubs, Hot and Cold Water in Rooms, Etc.

Meals 50 cents, $1.50 to $3.00 per day

COLFAX, CAL., April 17th 1906

Mr. Fred Searls

Dear Sir,

Should Dunlap and Simonds approach you and a settlement is agreeable, I only want to add the following to the general proposition which I submitted to you namely that a survey shall first be made to establish the exact location of the croppings of the Secompton ledge, and those of the Norton ledge & to establish their identity.

Mr. Norton is working cheerfully and will at once notify you of any opposition. I shall be back the 23 or 25th not later. With regards, sincerely yours,

Karl Howard

This letter, on Gillen Hotel stationery and dated April 17, 1906, confirms a previously negotiated agreement, emphasizing the proviso that a survey be conducted. The business card information at the top of the stationery boasts the Gillen Hotel's features, including electric lights, hot and cold water, and a porcelain bathtub in the individual rooms. Meals were only 50¢ apiece, or one could obtain a daily meal plan for $1.50 to $2. Steak dinners cost 75¢. (AS.)

The Gillen Hotel in Colfax (below) offered a grand dance hall known as Gillen Pavilion on the left side the hotel building, in front of the water tower. The structure also served as a venue for meetings, graduations, and other affairs. The back of this postcard has a message written in French. (CHM.)

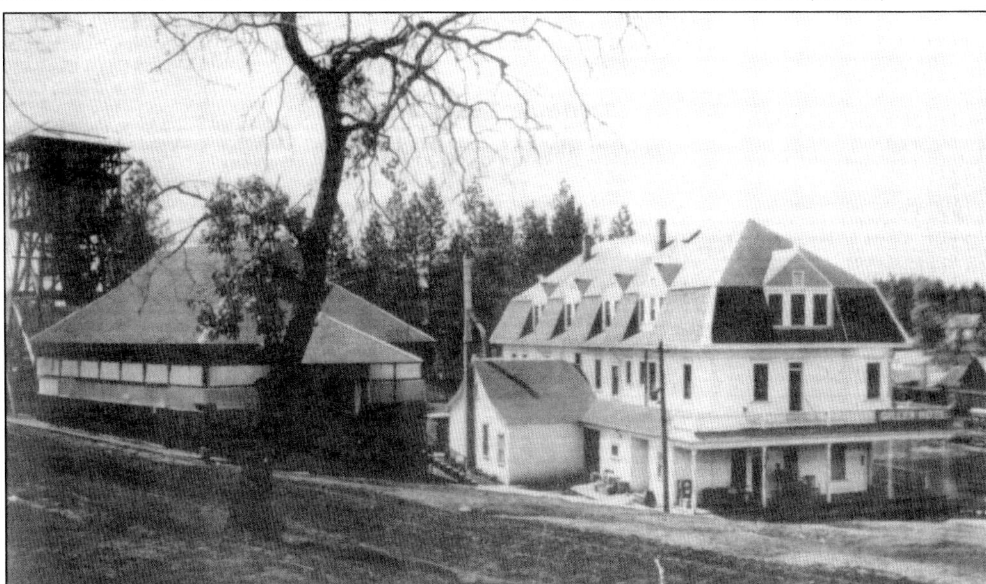

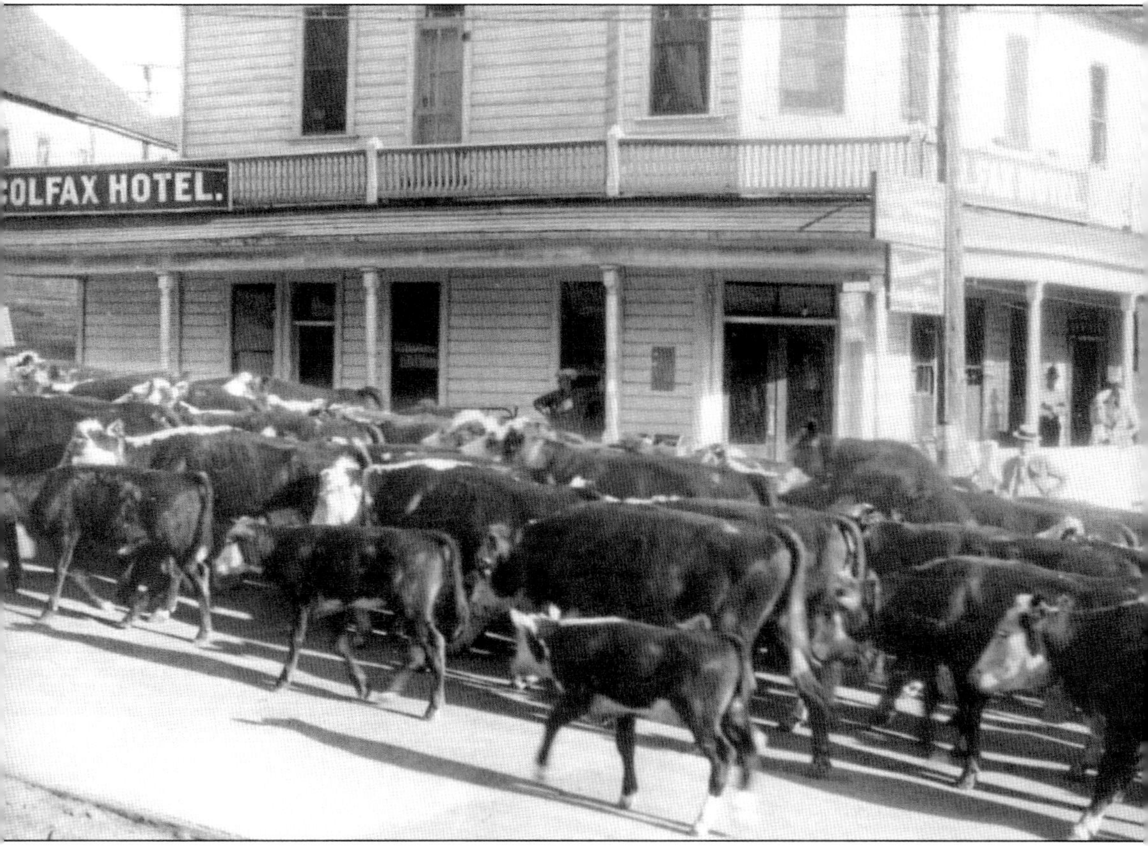

A cattle drive passes through the center of town directly in front of the Colfax Hotel. Daniel Gillen built the hotel at Grass Valley and Railroad Streets in 1897 after purchasing the land from D.A. Russell. The grand opening took place in 1903. Originally, the hotel offered 56 rooms, but then 11 rooms got assigned for workers and hotel management family only, leaving 45 rooms for visitors. The dining room accommodated 100 people. (CHM.)

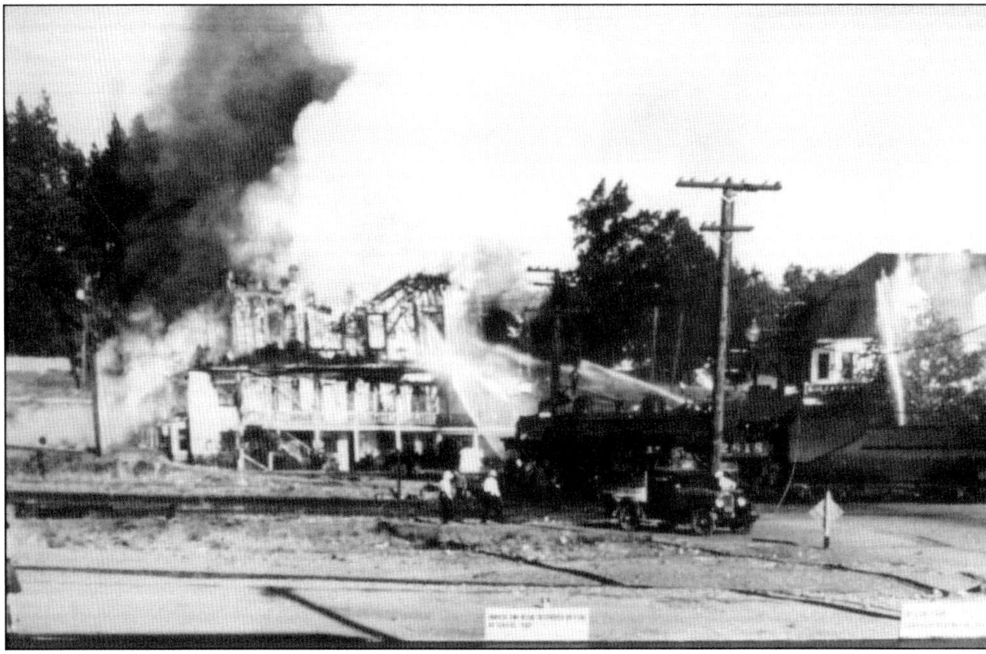

In 1905, Fred Marvin built the Marvin Inn as an annex to the Marvin Hotel. It was destroyed by fire in 1939. Charles McCormick and his wife ran the famous eating house for years. The Colfax Lions Club also met in the Marvin. (AS.)

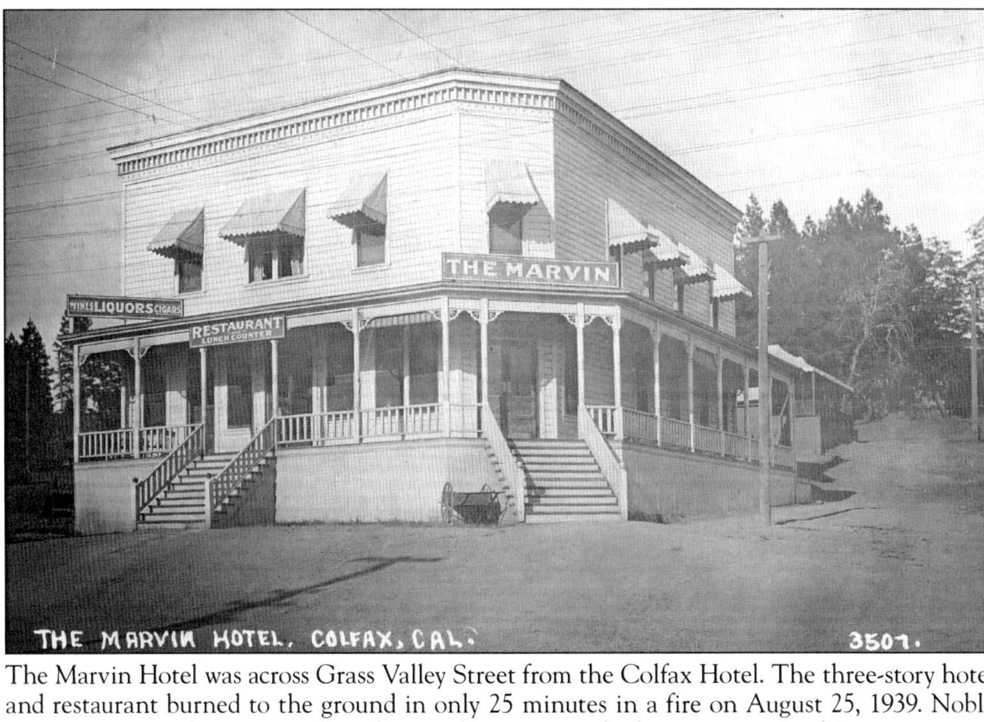

The Marvin Hotel was across Grass Valley Street from the Colfax Hotel. The three-story hotel and restaurant burned to the ground in only 25 minutes in a fire on August 25, 1939. Noble attempts to save the structure by volunteer firemen were futile. May Pepin and Dominio Curto had managed the inn. Thankfully, the fire did not spread to the Colfax Hotel. (CHM.)

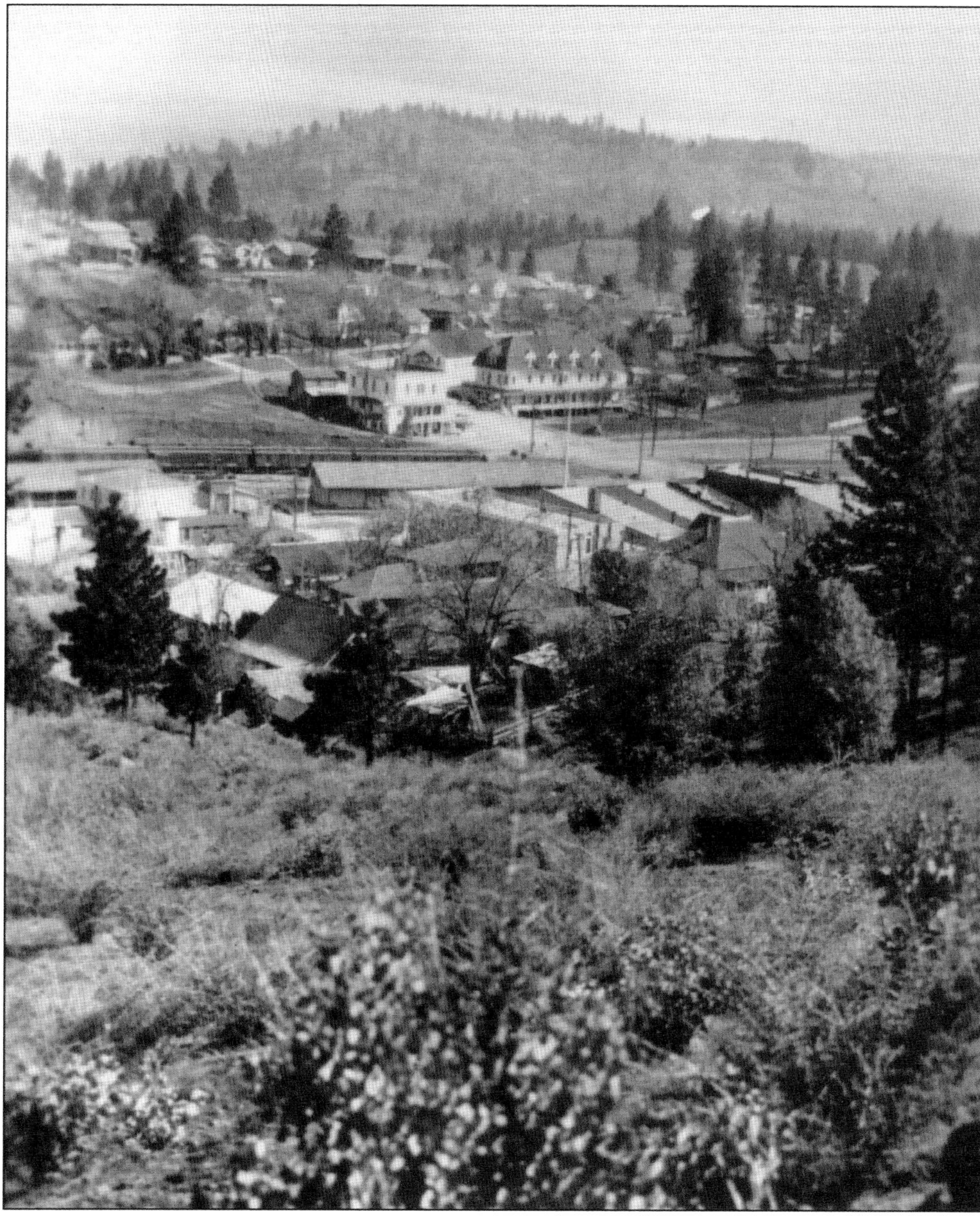

This scenic view of Colfax from the 1930s shows the town with its stretches of railroad tracks and a train sitting at the station. Midway up, on the other side of the railroad, are two large white buildings, the Marvin Hotel (left) and the Colfax Hotel. (CHM.)

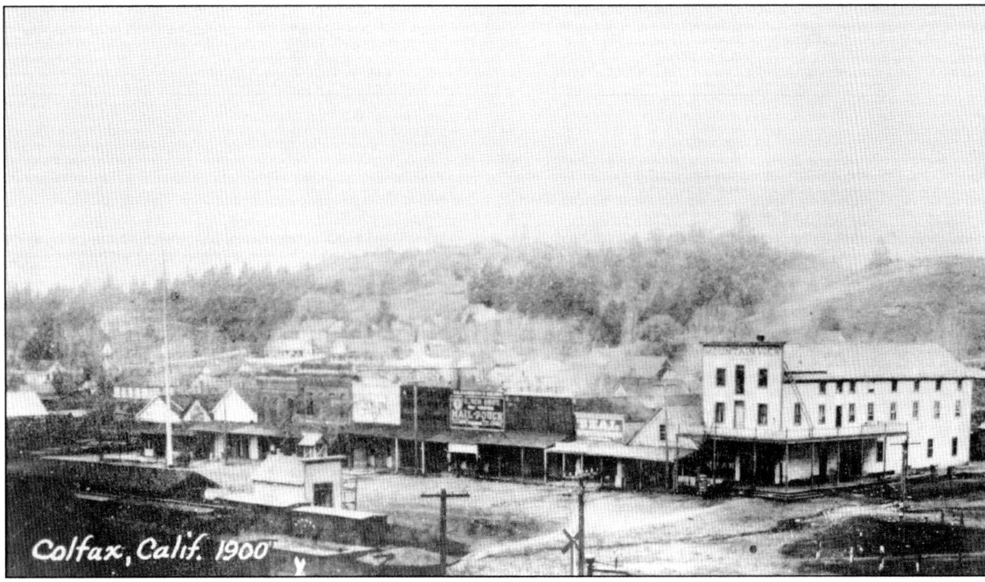

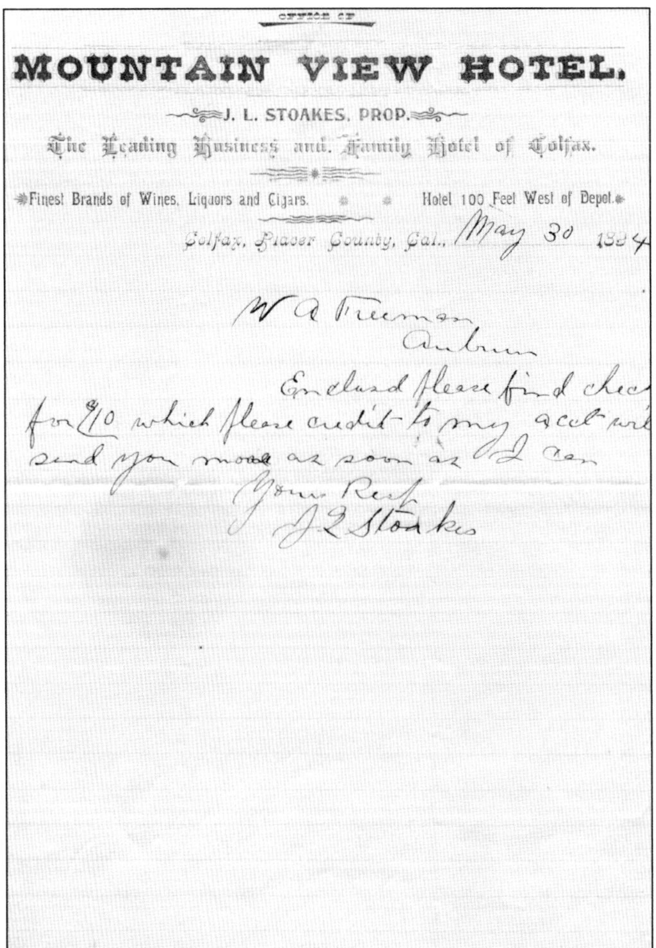

OFFICE OF

MOUNTAIN VIEW HOTEL.

J. L. STOAKES, PROP.

The Leading Business and Family Hotel of Colfax.

Finest Brands of Wines, Liquors and Cigars. Hotel 100 Feet West of Depot.

Colfax, Placer County, Cal., May 30 1894

W A Freeman
Auburn

Enclosed please find check for $10 which please credit to my acct will send you more as soon as I can

Yours Resp
J L Stokes

The Mountain View Hotel sat on the corner of Main and Depot Streets in Colfax (above). This 1900 photograph shows the building, which had the typical design of the era—a tall square facade rising above the railed balcony and masking the pitched roofline. The second story housed the hotel rooms, and a different business operated downstairs. The J.L. Stokes family owned the hotel until 1915. The structure burned down in 1983 and is now a vacant lot. (JLJ.)

This receipt, dated May 30, 1894, is from J.L. Stokes, the Mountain View Hotel proprietor, and directed to W.A. Freeman of Auburn. In the note, Stokes wrote that he included a check for $10 to be applied to his account as partial payment, with the promise for further payment in the future. It is not clear what sort of products or services Freeman supplied to the hotel. (AS.)

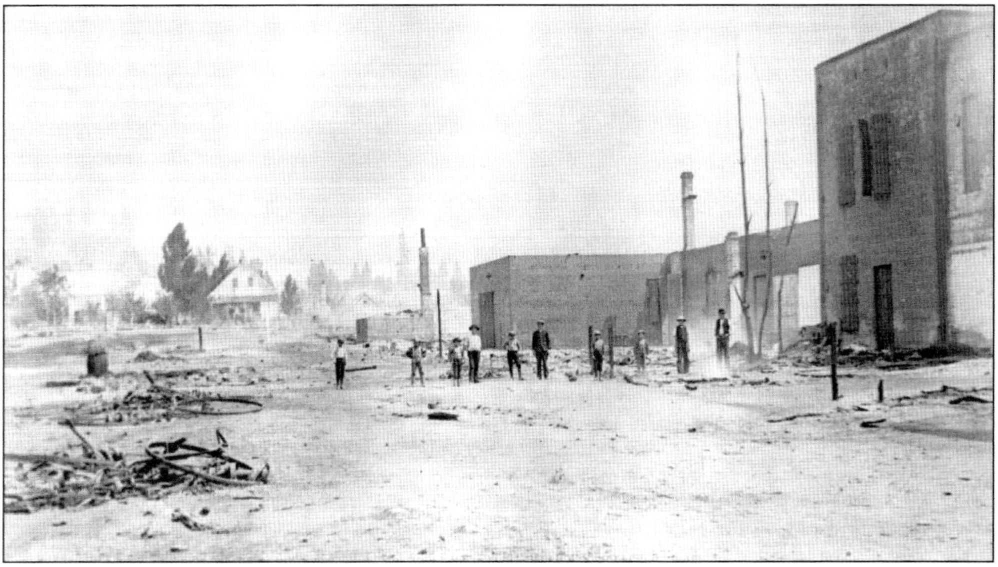

Fire was a constant problem in the early west. With entire towns built of wood, sometimes the whole town would burn—often more than once. Bucket brigades and water supplied from hoses were ineffective solutions once the flames took hold. In 1887, a major fire devastated Colfax. It was one of three major fires to decimate the town in the 1800s. This fire destroyed most of the town except for the brick buildings that were topped with two feet of dirt and a metal roof as fire protection. The surviving buildings included the Corner Market, the Keck Building, and the Masonic and Odd Fellows lodges. The only multi-building fire since 1900 burned eight structures in 1911. Local citizens examine the destruction caused by the 1887 Colfax fire. Frank Kuenzly Sr. stands on the far left. (AS.)

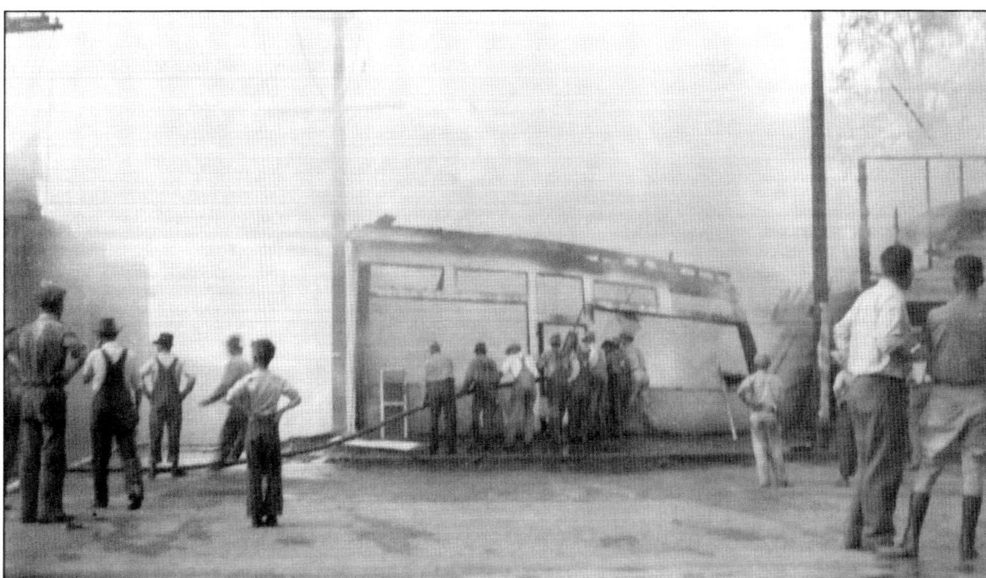

This photograph illustrates the manpower used in an attempt to douse the flames, as many men struggled to steady the hose. All was not lost in Colfax that day, but numerous structures required replacement. (AS.)

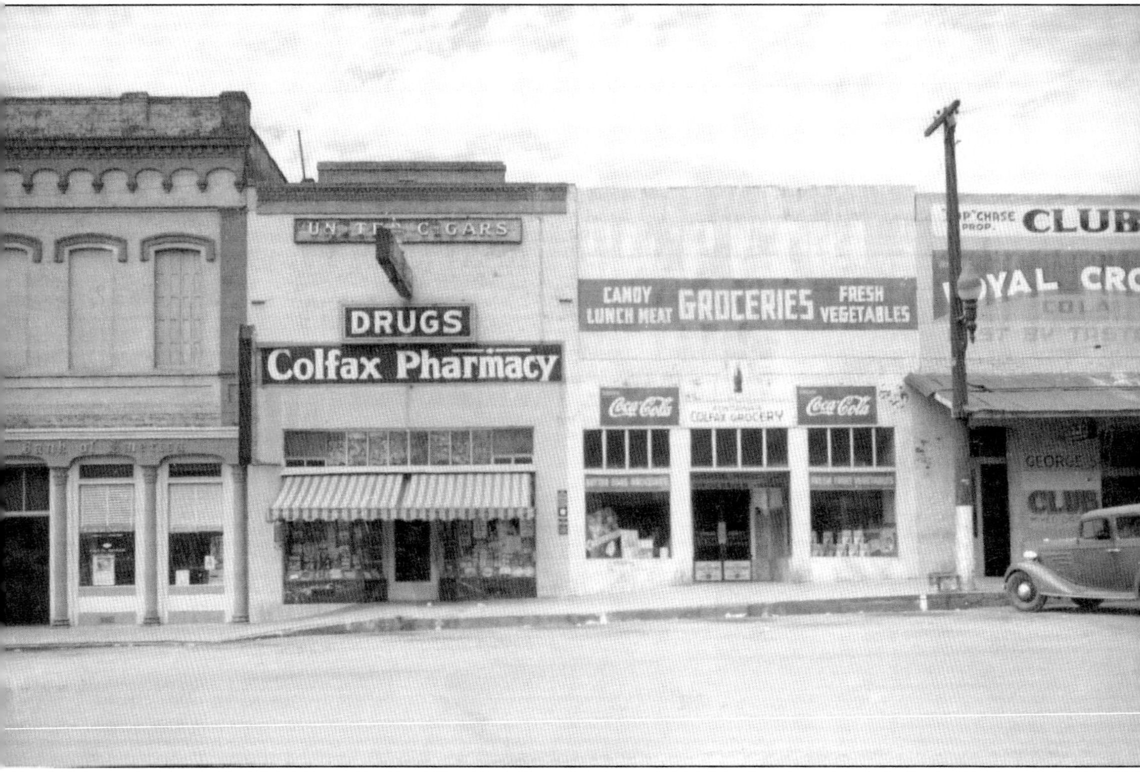

In 1870, Traphagen established the Colfax Pharmacy. He also served as the town dentist. The business changed owners several times, with John L. Butler purchasing the pharmacy from Traphagen in 1883 and moving it to 30 North Main Street. It remained for more than a century. Colfax Pharmacy was to become of the four oldest pharmacies in California. A post office also operated in the store. In 1912, the first telephone switchboard further enhanced the business. Sam Laing then bought it in 1929, adding the soda fountain that was later sold to the Colfax Hotel. During World War II, John Kaufman, the hotel proprietor, was forced to close the soda fountain for part of the war because one precious ingredient proved unavailable: chocolate. The pharmacy became the official dispatch location for gifts shipped to American soldiers overseas. In 1956, pharmacist Richard Wayland purchased it with two partners, Robert Kaufman, a local auto dealer, and Garvin Mitchell, the former manager of the Colfax Fruit Association. Wayland had worked as a stock boy in the local pharmacy beginning in grammar school. His family ran the business for 40 years, until 1996, when it was bought by Payless Drugs. (AS.)

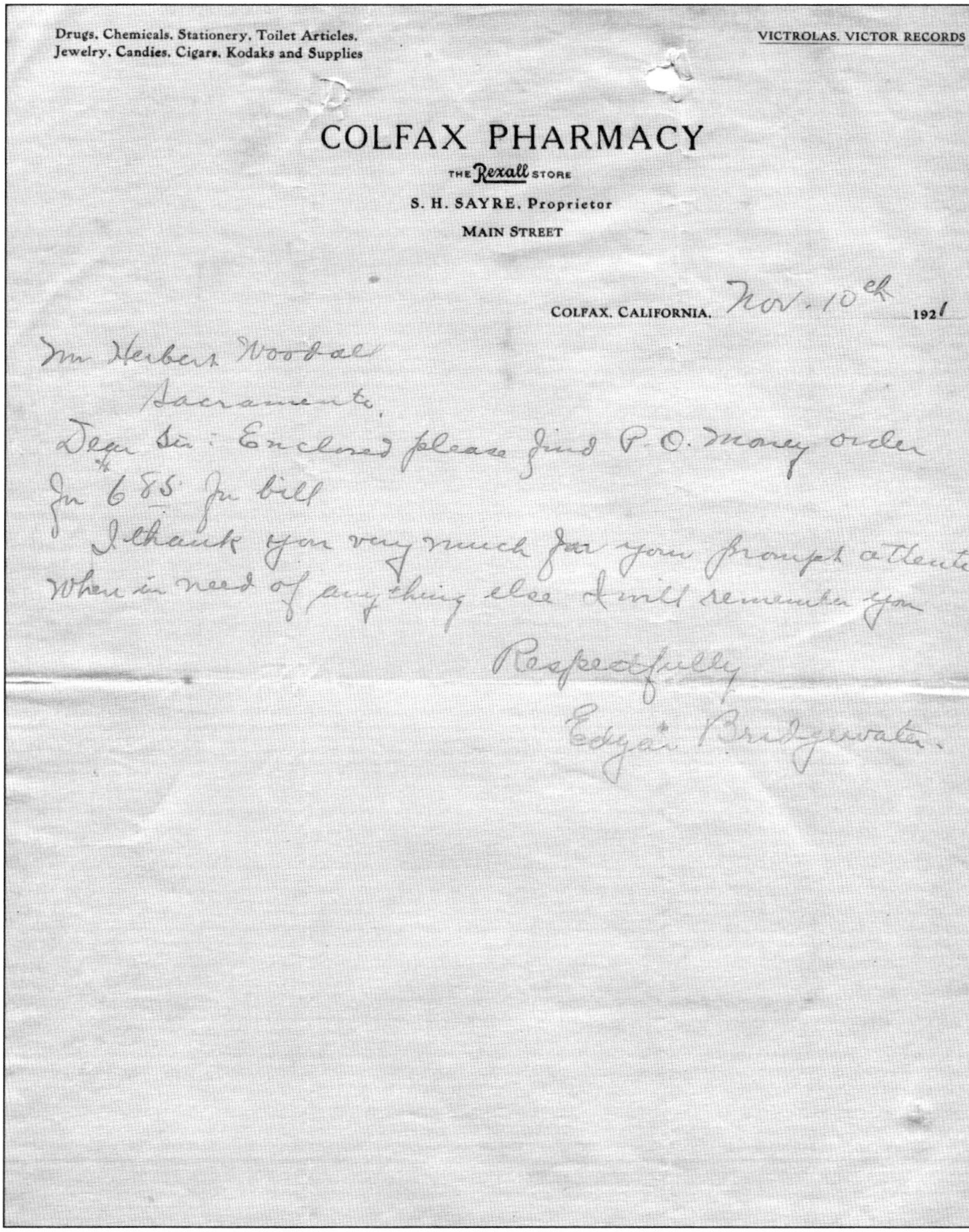
Drugs, Chemicals, Stationery, Toilet Articles,
Jewelry, Candies, Cigars, Kodaks and Supplies

VICTROLAS. VICTOR RECORDS

COLFAX PHARMACY

THE Rexall STORE

S. H. SAYRE, Proprietor

MAIN STREET

COLFAX, CALIFORNIA, Nov. 10th 1921

Mr Herbert Woodall
Sacramento,
Dear Sir: Enclosed please find P.O. money order for 6.85 for bill
I thank you very much for your prompt attenti
When in need of anything else I will remember you
Respectfully
Edgar Bridgewater.

This Colfax Pharmacy receipt, dated November 10, 1921, was written by Edgar Bridgewater to Herbert Woodall and applies a money order in the amount of $6.85 as payment on the account. He expressed sincere appreciation to Herbert Woodall for the pharmacy's quick response. The business delivered prescriptions, oxygen, and home health supplies to three nursing homes. It also made deliveries as far away as Rancho Seco in Sacramento, Scotts Flat in Nevada City, and Rollins Lake just north of Colfax, off Highway 174 on the way to Grass Valley. As the Yuba County and Placer County water agencies engaged in new projects, the Colfax Pharmacy brought first-aid supplies to their sites. Over its history, the store sold everything from Victrola phonograph records and sheet music to tobacco, liquor, and ice cream, along with sundries and prescription drugs. (AS.)

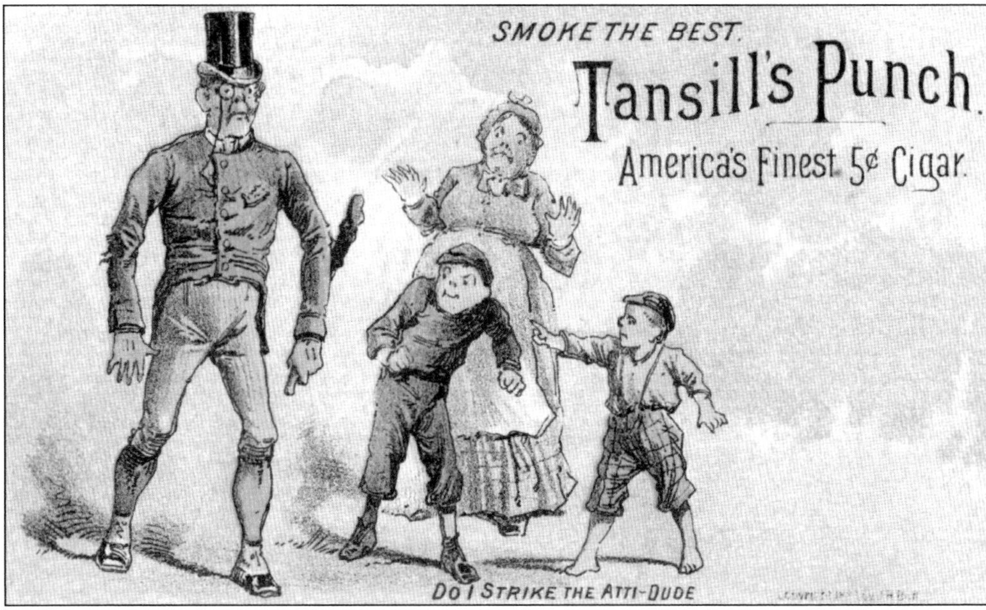

These whimsical advertisements promote the attributes of cigar smoking while touting Tansill's Punch as the best of the 5¢ cigars. Robert Weems Tansill began his Tansill's Punch knockoff cigar business in 1862. The cigars derived the name from the 17th-century "Punch and Judy" show in Great Britain. The advertisement below, while definitely not politically correct or appropriate, reflects the stereotypes of another era. Tansill became renowned for his creative advertising and promotion techniques. The drawings on the advertisements are dated by the original 1882 copyright of J.H. Bufford. The reverse side of the cards indicates John Butler as an agent for Tansill and a dealer of everything else, from pharmaceuticals to perfumes. His place of business also served as the post office at that time. A New Zealand pharmaceutical publication described the cigars thusly: "When a man smokes one of these cigars, he walks on air." (Both, AS.)

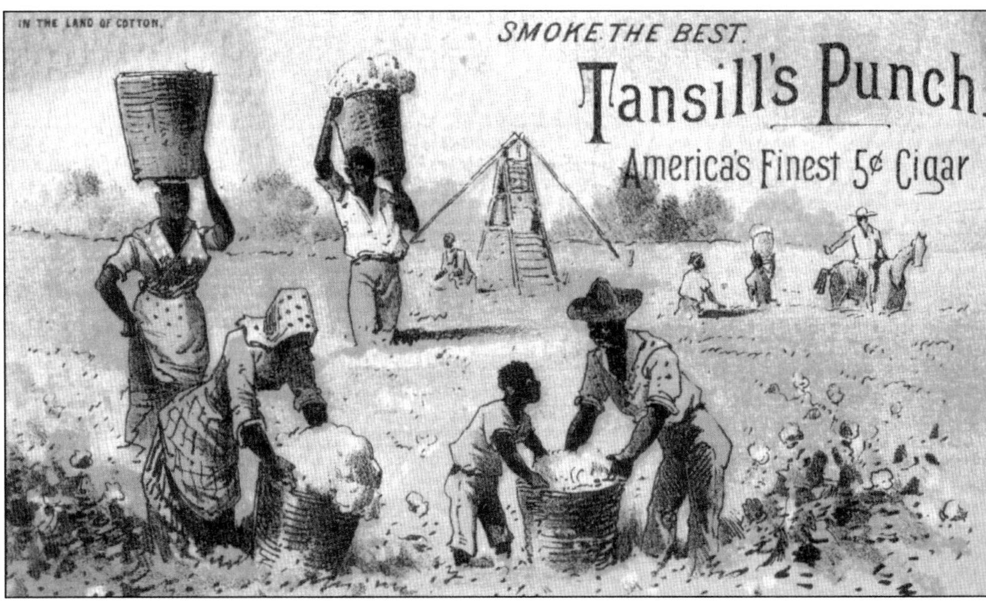

John Butler, the pharmacy proprietor and local Colfax cigar dealer, posed for this formal photograph. One of his advertisements stated, "Prescriptions carefully compounded. Musical instruments of all descriptions. Orders taken for any kind of sheet music and delivered on short notice. Try our ice cold soda and fresh roasted peanuts." (CHM.)

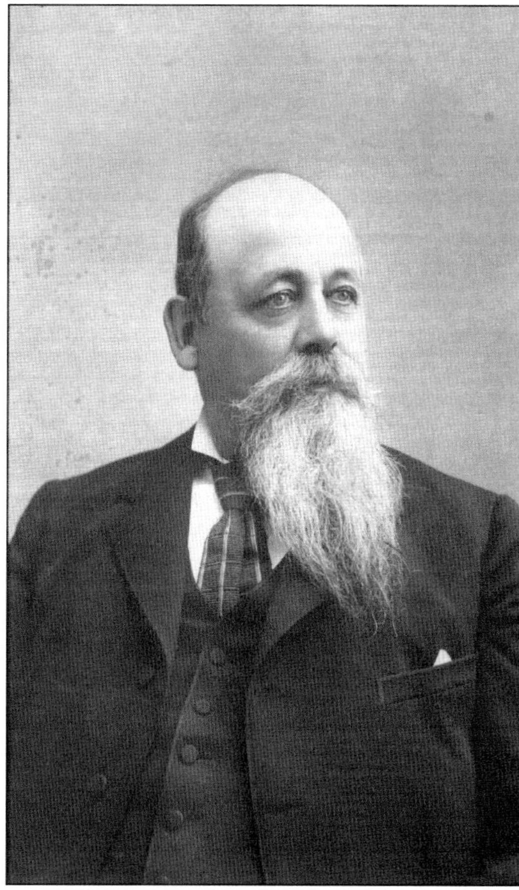

The Lobner store (below) served the community as a general merchandise store. Bins of local fresh fruit set in the front tempted passersby and serious shoppers to stop in and make purchases. Goods lined both sides, almost reaching the ceiling and requiring ladders to grab some of the highest items. The contractors built every store with a similar design—long and narrow with very high ceilings. Unusual light fixtures hung prominently in the center of the establishment. (CHM.)

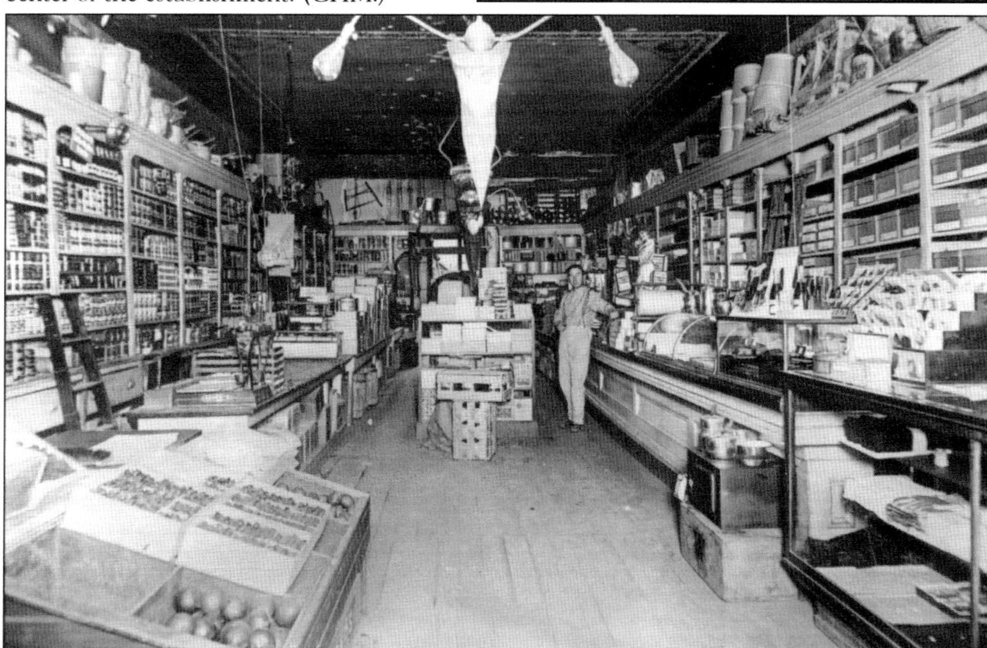

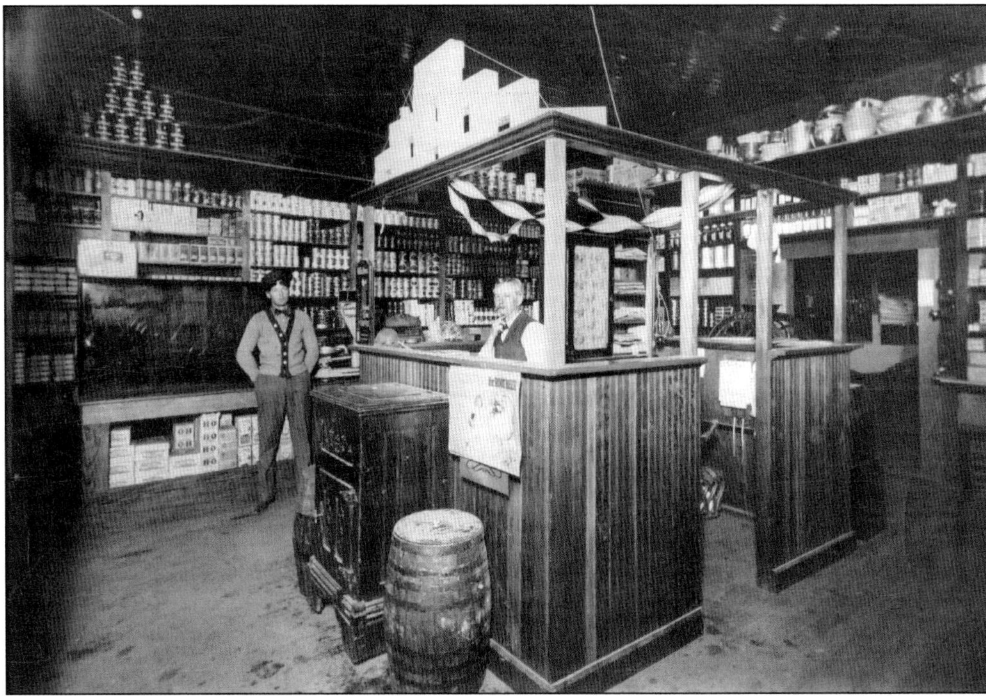

The interiors of retail stores like the one above show the reality of merchandising in 1890. The place overflows with goods in cans, barrels, and crates. This photograph shows that the store was equipped with a woodstove, a necessity for heat in the winter. A barrel sits in front of the cashier, perhaps filled with flour, grains, or even pickles. There is a faint sign promoting root beer, a popular drink at the time, and the upper shelf is lined with essential cooking wares. (CHM.)

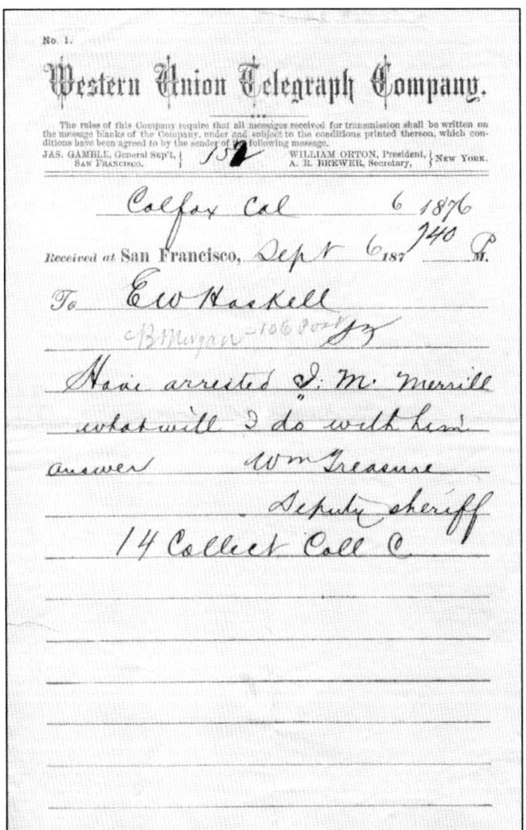

No. 1.

Western Union Telegraph Company.

The rules of this Company require that all messages received for transmission shall be written on the message blanks of the Company, under and subject to the conditions printed thereon, which conditions have been agreed to by the sender of the following message.

JAS. GAMBLE, General Sup't, San Francisco. 152 WILLIAM ORTON, President, A. R. BREWER, Secretary, New York.

Colfox Cal 6 1876

Received at San Francisco, Sept 6 187 740 P.M.

To E W Haskell

[illegible]

Have arrested I. M. Merrill
what will I do with him
answer Wm Treasure
Deputy sheriff
14 Collect Coll C

Western Union provided quick communication via telegraph before the telephone was invented. In September 1876, deputy sheriff William Treasure telegraphed this message to E.W. Haskell, requesting advice as to what he should do with his new prisoner, I.M. Merrill. The offense committed by Merrill is unknown. (AS.)

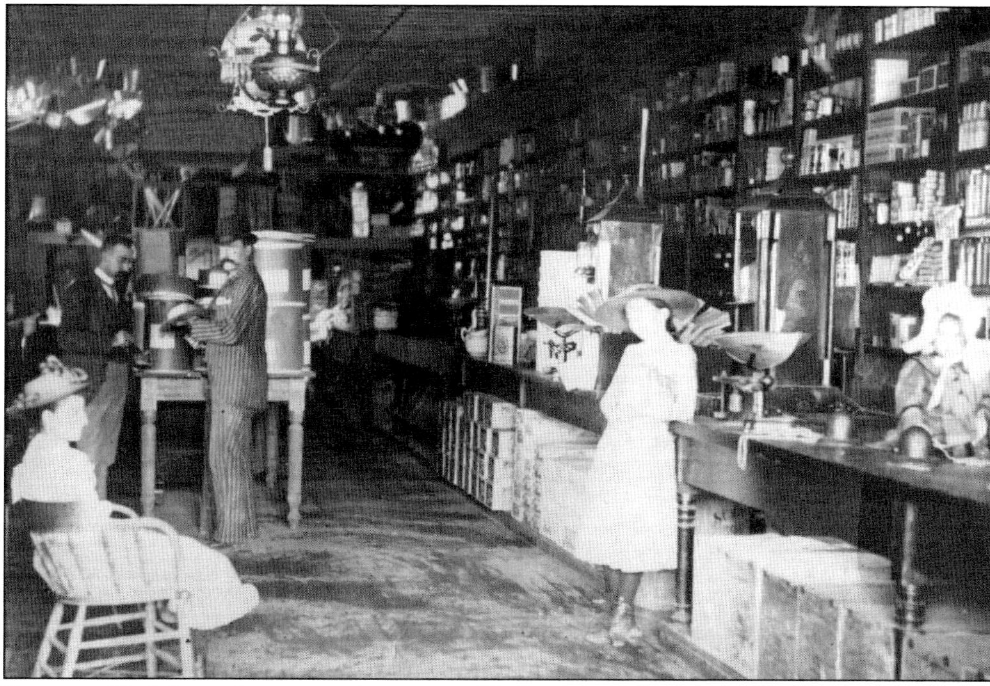

Benjamin Keck began his grocery business in Illinoistown with just a sack of flour. In 1867, after the Central Pacific completed construction of the transcontinental railroad, Keck moved to Colfax, where he ran a harness shop in town with seven harness-makers. Keck's granddaughter Alberta Cross commented in an oral history that the grocery wagon delivered groceries in the morning, but if there was a funeral that day, a black drape covered the sign and the wagon was used as a hearse in the afternoon. On occasion, the wagon even hauled drunks to jail. (CHM.)

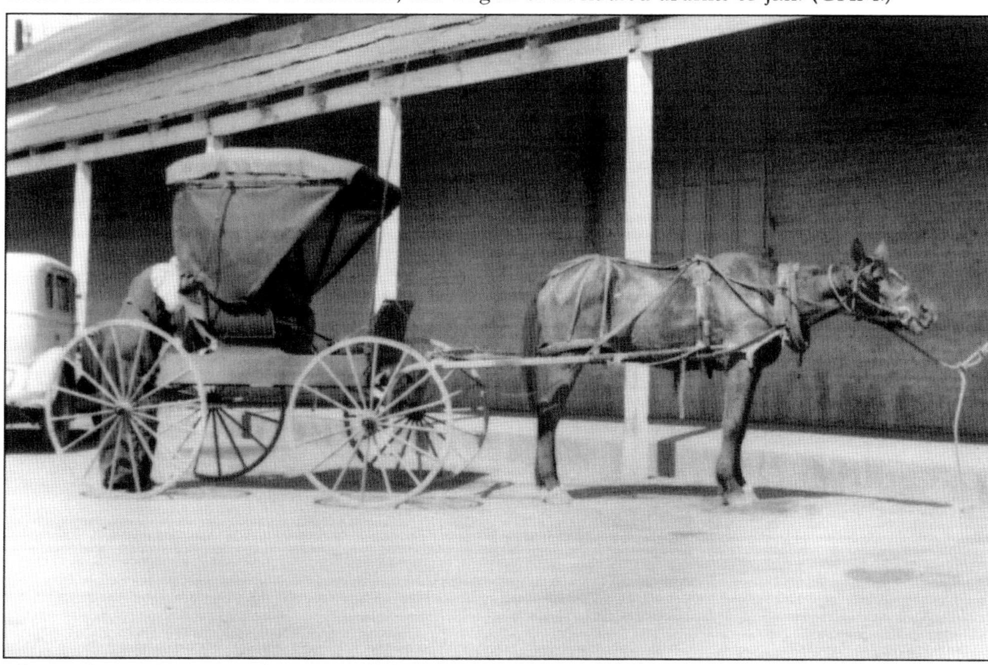

This is the multipurpose buggy used for groceries, corpses, and drunks. (CHM.)

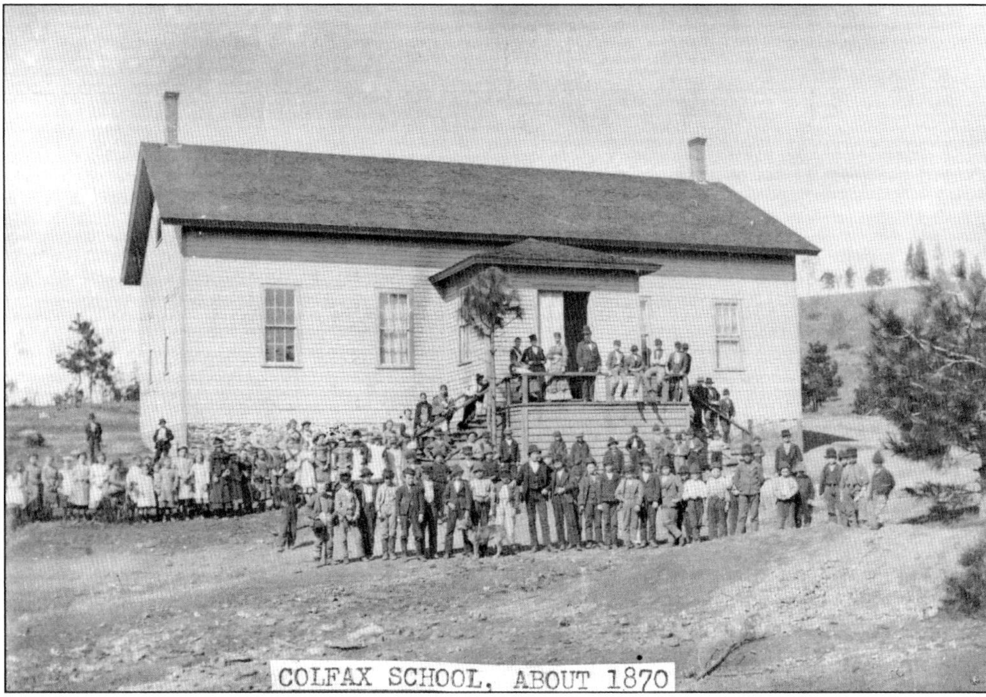

The first school for the local children was built in Auburn in 1852. Illinoistown was next, building a schoolhouse of its own. This photograph of the Colfax School, dated about 1870, shows all the children and teachers gathered in front the school building. Some are at attention while others are more casual. Every man or boy wears a customary hat. (PCM.)

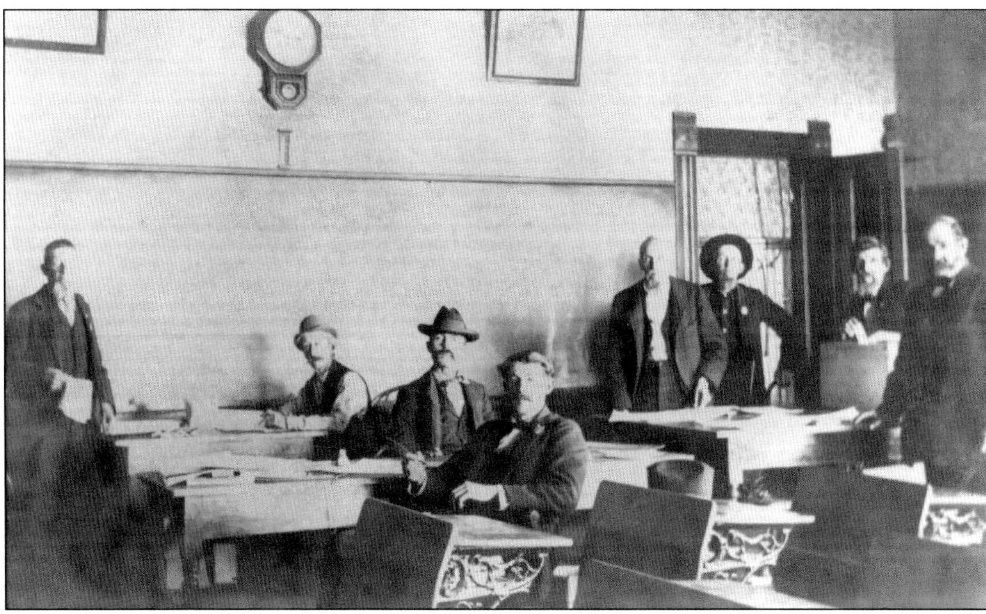

The board of education met in the classrooms of the schoolhouse. The old-fashioned desks attached the table of one desk to the chair in front of it. Ornate ironwork decorated the sides. This particular board meeting was held in 1898. (CHM.)

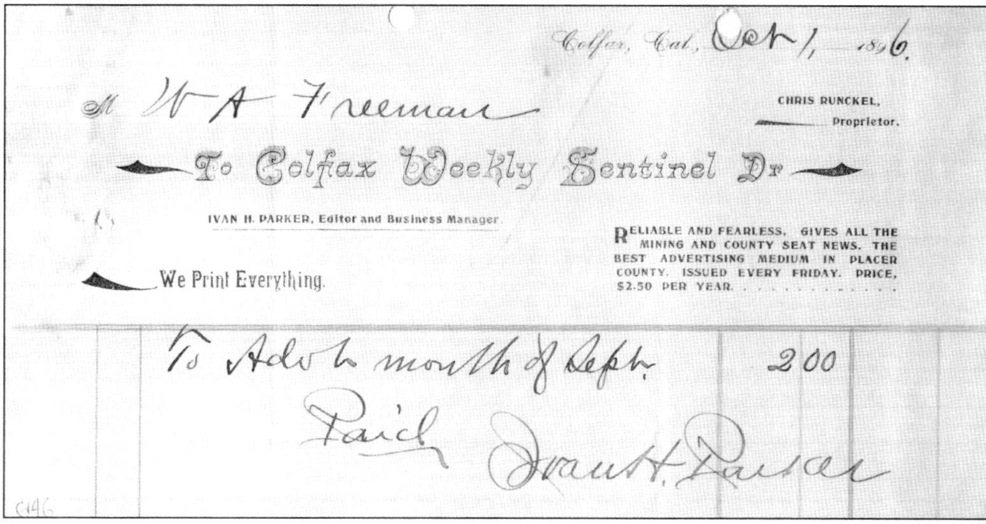

Colfax, Cal., Oct 1, 1896.

M W A Freeman

CHRIS RUNCKEL, Proprietor.

To Colfax Weekly Sentinel Dr.

IVAN H. PARKER, Editor and Business Manager.

RELIABLE AND FEARLESS. GIVES ALL THE MINING AND COUNTY SEAT NEWS. THE BEST ADVERTISING MEDIUM IN PLACER COUNTY. ISSUED EVERY FRIDAY. PRICE, $2.50 PER YEAR.

We Print Everything.

To Adv. month of Sept. 2 00

Paid Ivan H. Parker

William T. Calkins, publisher, and James C. Tyrell, managing editor, established the *Colfax Weekly Sentinel* in August 1890. The paper was sold and resold several times until it went into foreclosure in March 1908. The *Colfax Record* began publication on November 26, 1908, and still turns out local news each week. It is one of the oldest weekly newspapers in the state of California. This October 1, 1896, receipt advances payment of $2 for an advertisement in the *Weekly Sentinel*. (AS.)

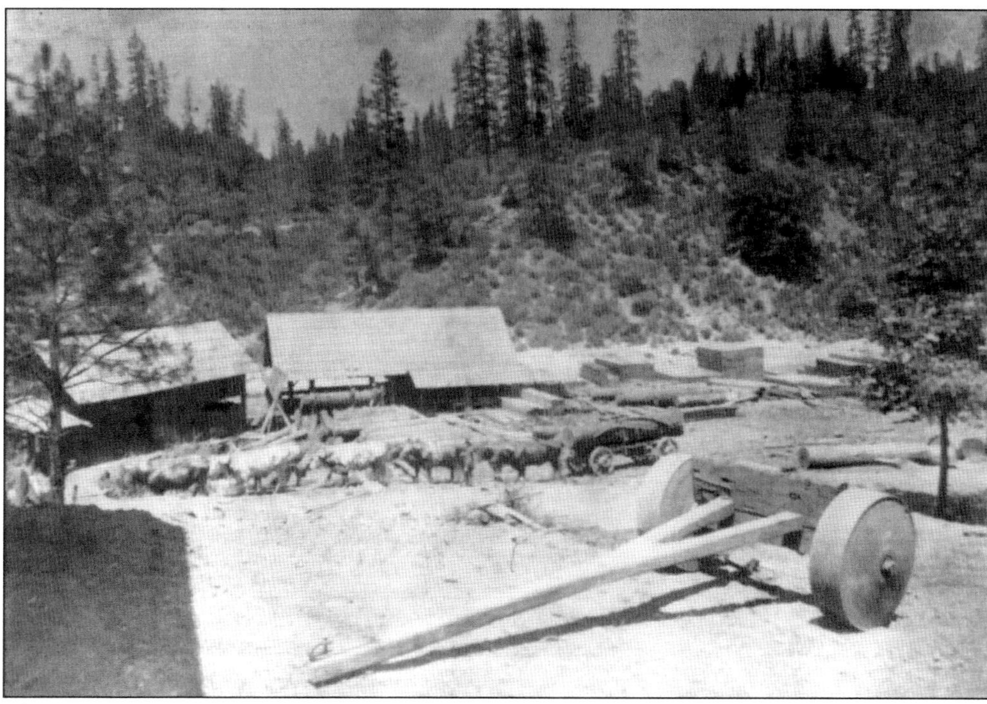

This sawmill operated on the Bear River near Colfax. According to statistics for the area, more than 44 million board feet and more than one million lineal feet of round timber had been sawed by 1873. Large logs were transported on wagons pulled by teams of oxen. Lumbering became big business in the foothills when the transcontinental railroad was constructed, when the towns grew, and then again when the underground mines needed to be fortified. (LOC.)

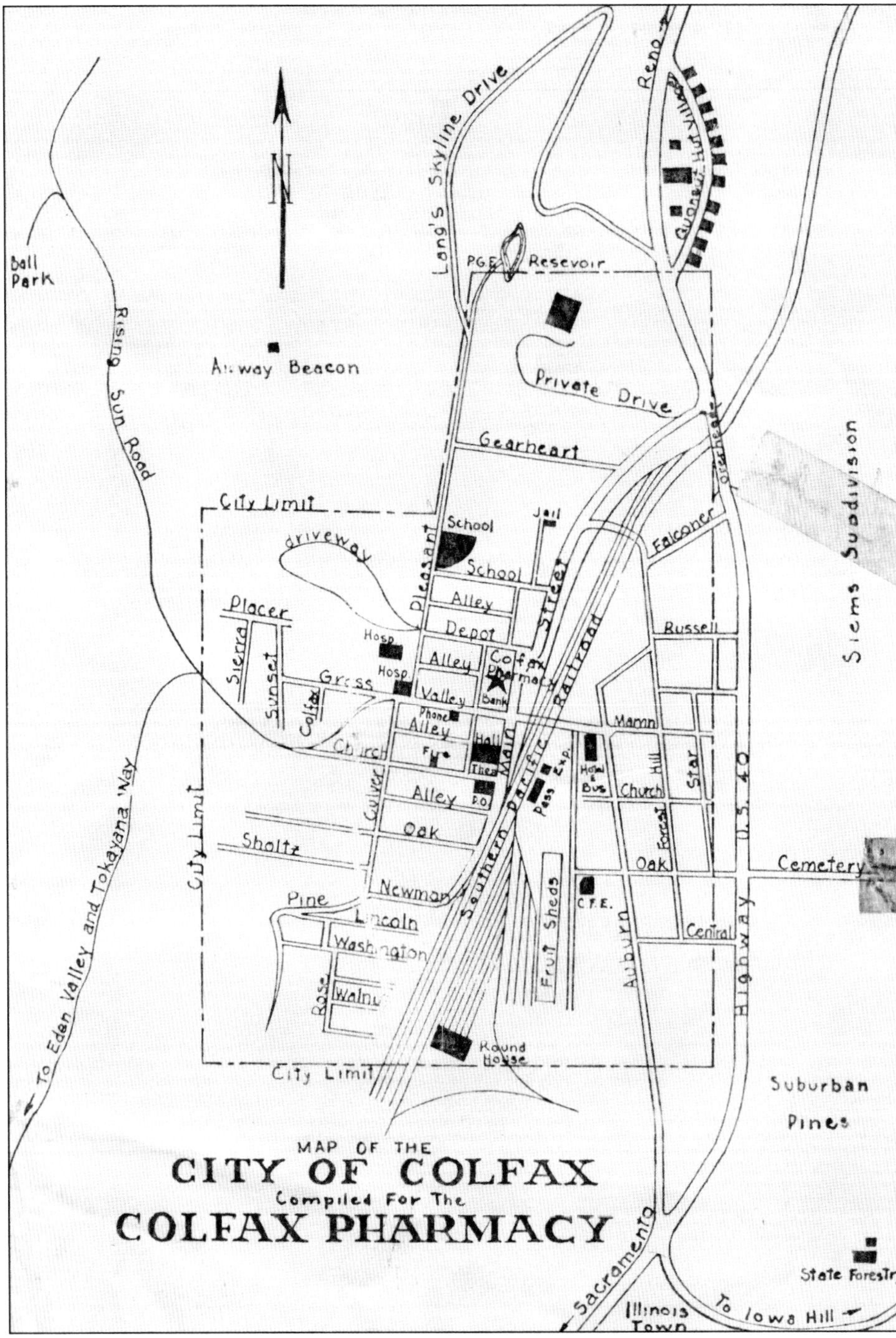

This street map of Colfax shows the layout of the current town. Many buildings have changed owners, tenants, and signs, but the streets remain. The map helps give perspective for the images in the next chapter. Today, there are numerous plaques on the fronts of the historic buildings honoring the businesses that originated there. The town has also placed other markers at significant locations, acknowledging historic contributions with a total of more than two dozen historic plaques. (CHM.)

Five

Time Marches On

In the early 1900s, even more change came to Colfax, the state of California, and the world. Railroads, which previously transported freight and passengers, soon became essential to World War I efforts, moving troops and supplies. Automobiles and airplanes offered new modes of transportation for individuals, businesses, and the military. Roads needed to be added and improved to meet the demand.

Communities in the foothills became stronger, with a growing number of families and children. New forms of entertainment designed to engage all members of the community evolved. Baseball teams, parades, and celebrations drew people together. The addition of electricity and telephones altered the way people lived and communicated, and the Industrial Revolution offered many new conveniences and efficiencies for society.

Although gold mining by corporations still turned a profit in the Sierra foothills, fewer individuals found it rewarding. The foothills did not suffer the extreme hardships of the Depression after the 1929 stock market crash because mining continued, creating income and employment. Gold once again became essential to finance World War II, but by the time the United States entered the war, the government required the mines to shut down so that resources, raw materials, and people could be diverted from mining to wartime manufacturing and transportation. Some mines reopened after the war, but the output no longer compensated for the costs. Towns like Colfax became more reliant on the railroad, lumber, construction trades, agriculture, and tourism. The attraction of the rich history of the Colfax area, along with the charm and convenience of traveling by train, continue to draw curious visitors from near and far.

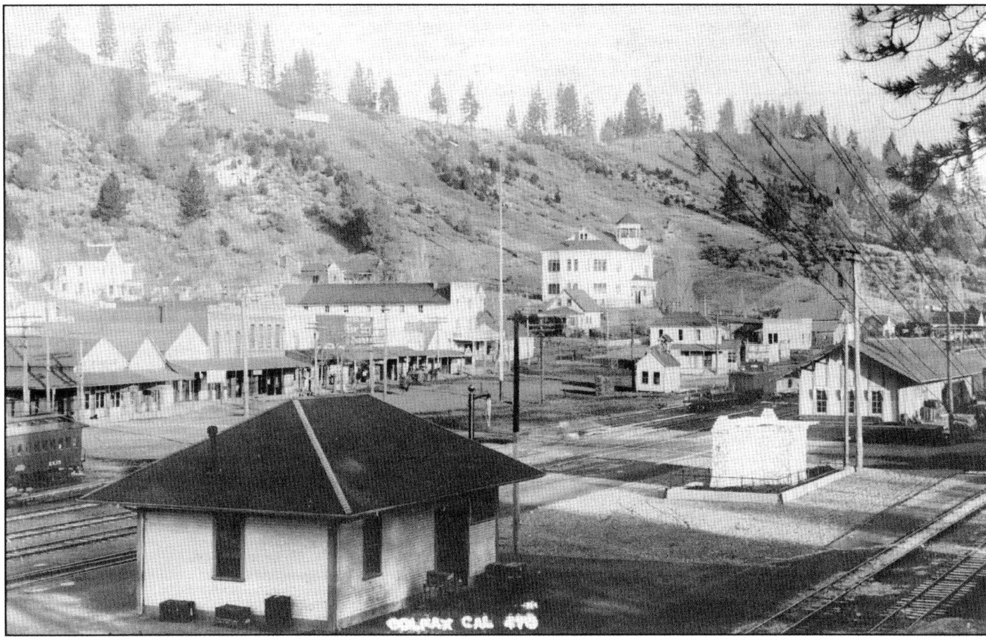

This 1900 postcard illustrates the business section of town on the left, but it also shows a clear view of the residential area behind Main Street going up the hill. The Kuenzly-Lobner house, a large white Victorian, stands out in the center right, perfectly situated to observe the town and the railroad activities. The large white building on the right behind downtown was the school building that opened in 1894 and burned in 1912. The Nevada County Quartz Monument (center right) sat between the two railroad depots. (PCM.)

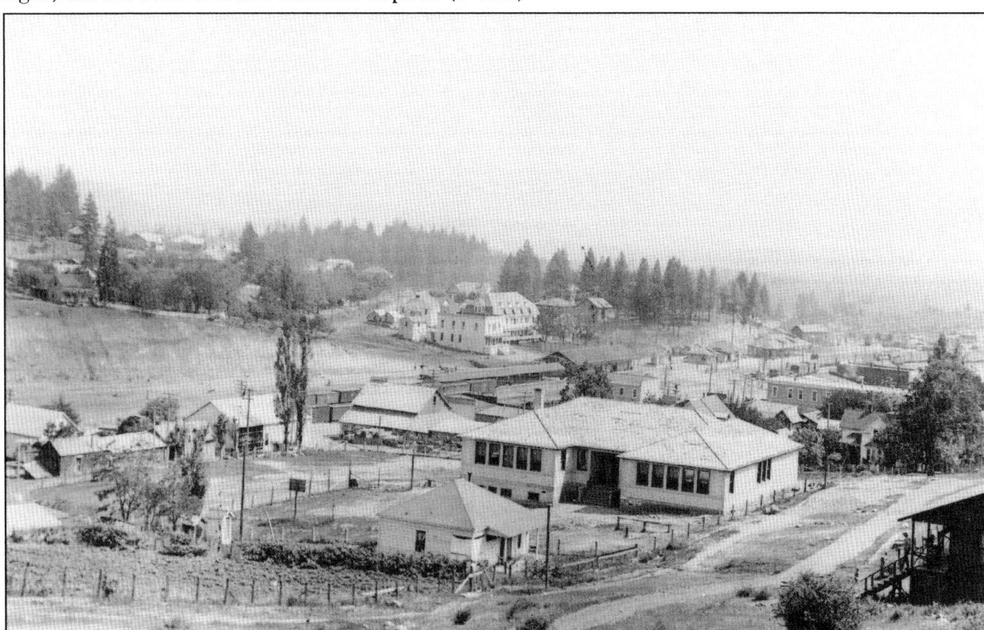

A new Colfax Grammar School opened in 1913. It is the large white building with a recessed entry toward the front of the photograph. The town built it to replace the one that burned the year before. The three-story white building toward the back in the center is the Colfax Hotel. (PCM.)

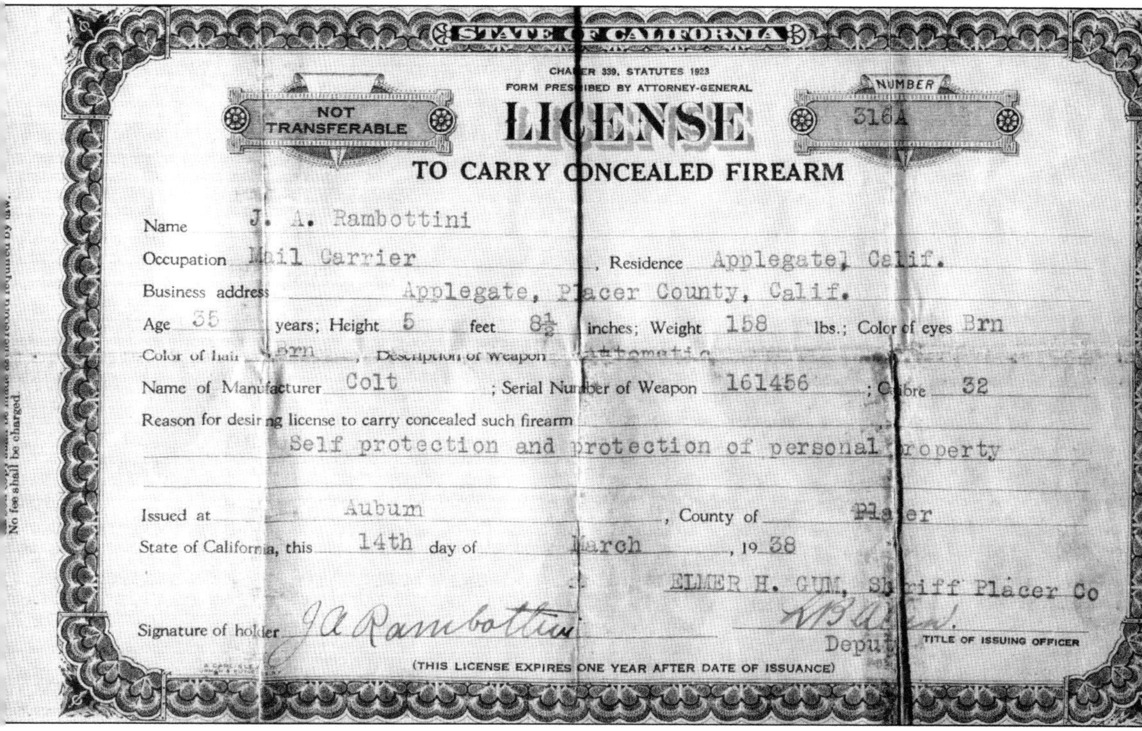

STATE OF CALIFORNIA

CHAPTER 339, STATUTES 1923
FORM PRESCRIBED BY ATTORNEY-GENERAL

NOT TRANSFERABLE

LICENSE

NUMBER 316A

TO CARRY CONCEALED FIREARM

Name J. A. Rambottini

Occupation Mail Carrier, Residence Applegate, Calif.

Business address Applegate, Placer County, Calif.

Age 35 years; Height 5 feet 8½ inches; Weight 158 lbs.; Color of eyes Brn

Color of hair Brn, Description of Weapon Automatic

Name of Manufacturer Colt; Serial Number of Weapon 161456; Calibre 32

Reason for desiring license to carry concealed such firearm
Self protection and protection of personal property

Issued at Auburn, County of Placer

State of California, this 14th day of March, 1938

ELMER H. GUM, Sheriff Placer Co

Signature of holder J A Rambottini

Deputy

TITLE OF ISSUING OFFICER

(THIS LICENSE EXPIRES ONE YEAR AFTER DATE OF ISSUANCE)

No fee shall be charged

Pony Express riders, railroad engineers, and many other citizens in the early west carried guns as protection for themselves and their possessions. Stagecoach robberies became a problem for drivers on the lonesome roads and for Southern Pacific engineers over the summit route between the Nevada state line and Colfax. This firearm license from 1938 was issued for a concealed Colt automatic to J.A. Rombottini of Applegate. Elmer H. Gum, the Placer County sheriff, approved the license. (CHM.)

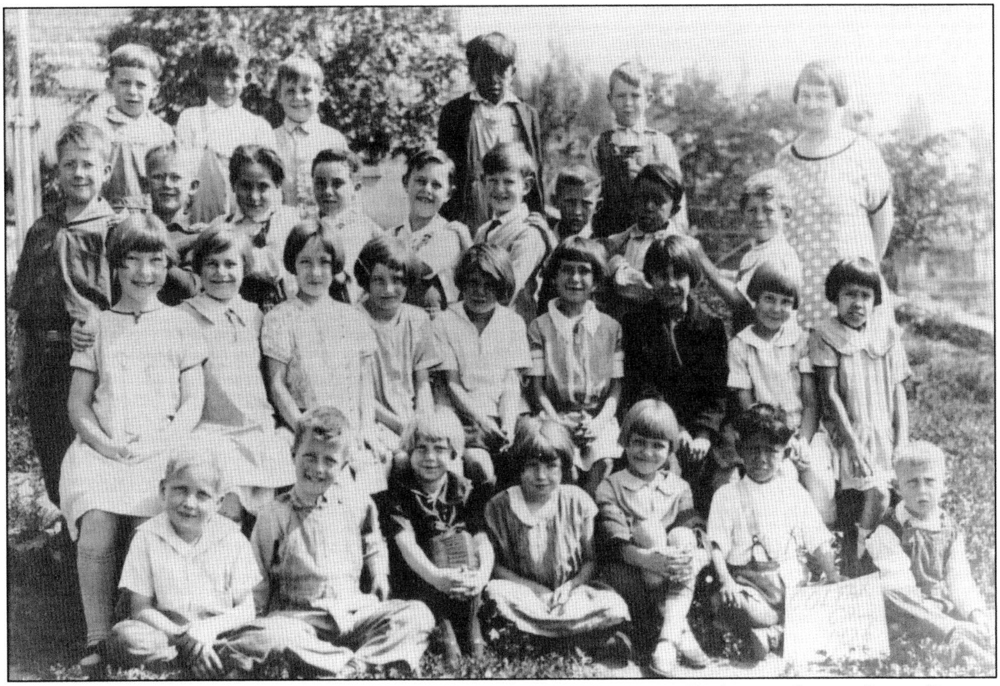

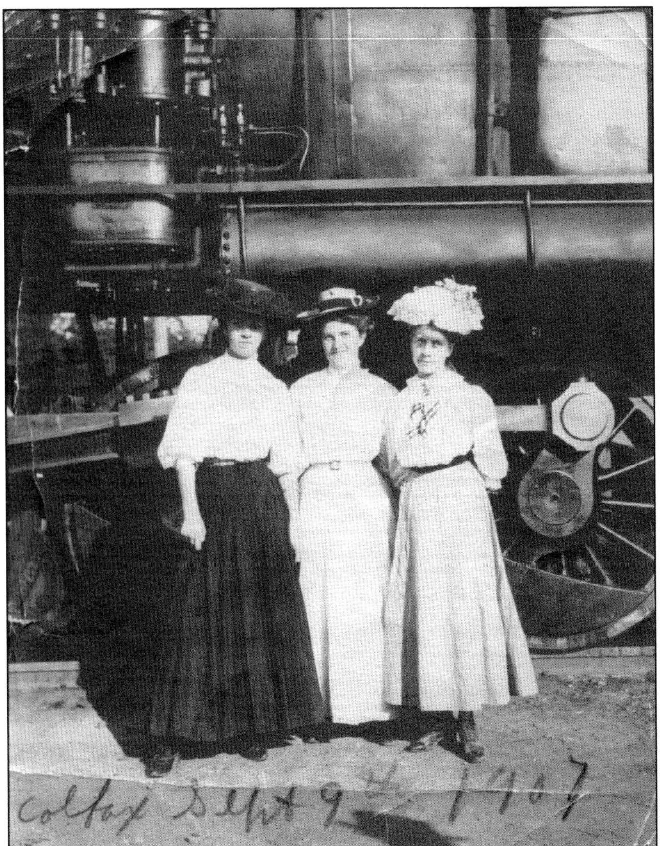

Fashions changed rapidly, just as they do today. The 1926 class photograph above shows the 30 children and their teacher in casual attire. All of the girls have their hair bobbed. During the Gold Rush, children had dressed differently and were fewer in number. In 1856, a total of 113 children attended school in Illinoistown. By 1881, there were 203 in Colfax. (CHM.)

This photograph of three pretty ladies in front of the locomotive in Colfax is a fine illustration of the fashions of the day. Note the grand high hats, modest necklines, and long flowing skirts that covered the tops of their button-down shoes. The photograph is dated September 9, 1907. (AS.)

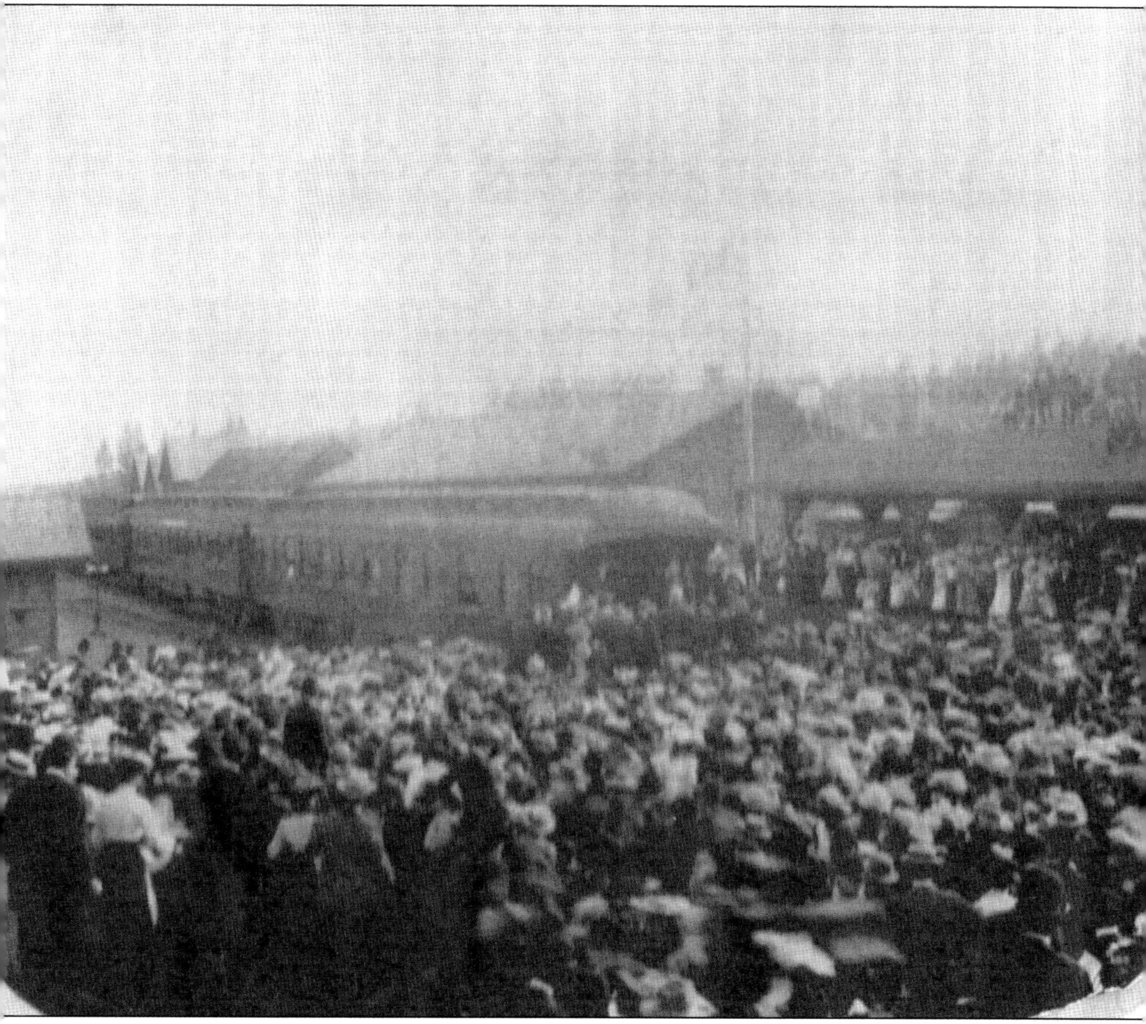

On May 19, 1903, Pres. Theodore Roosevelt arrived in Colfax by train for a brief stop and a few words. A crowd of about 3,000 spectators converged on the town for the event. The Nevada County Narrow Gauge advertised a "President's Special" so folks from Nevada County could come hear the president's speech. The cost was 75¢ for gentlemen and 50¢ for women and children. President Roosevelt said, "It is my greatest pleasure to have caught even a glimpse of the miners here." He added, "You in your lives here have done the things which it makes all of us proud as Americans to have done. We do not believe here in this republic in the men who seek only the life of ease, the life absent of effort. We believe in the men who face toil, who face risk, who dare, who do and who triumph because they have done it." (CHM.)

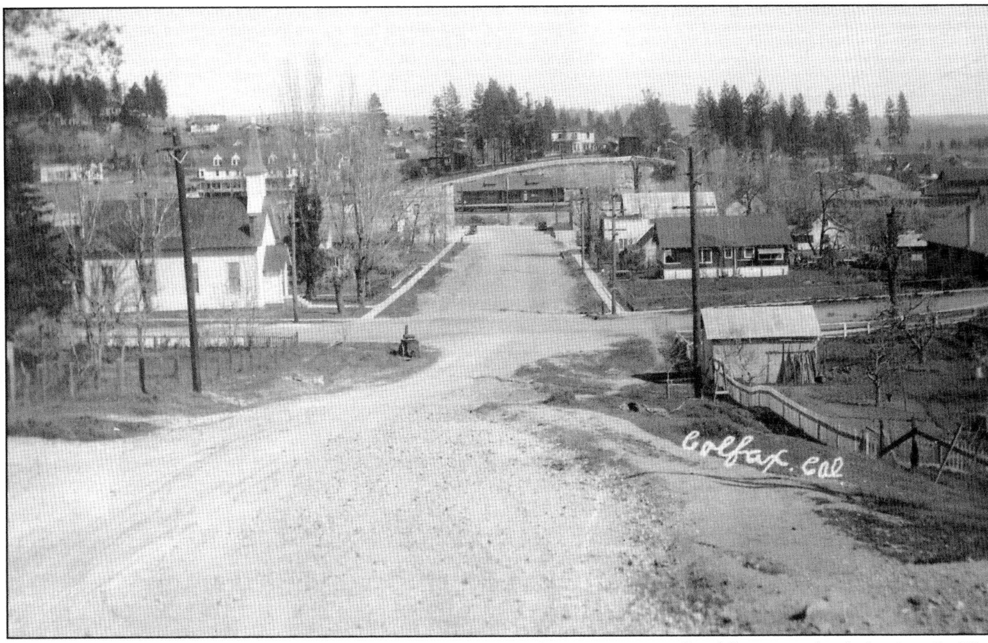

This view looks down Church Street and highlights the prominent building on the left, the Colfax Methodist Episcopal Church. Established in 1874 by Rev. I.J. Ross, the pastor of the Iowa Hill Church, this is the oldest church still in use in Colfax today. Just past the steeple, a few blocks back, stands the Colfax Hotel. (JLJ.)

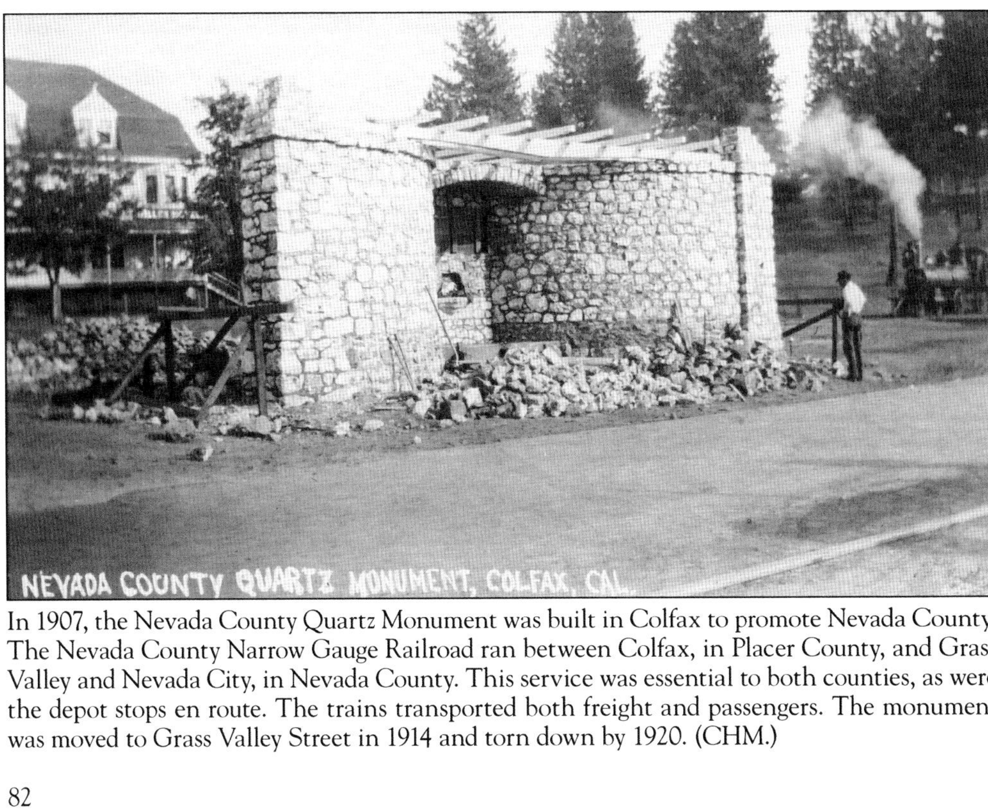

In 1907, the Nevada County Quartz Monument was built in Colfax to promote Nevada County. The Nevada County Narrow Gauge Railroad ran between Colfax, in Placer County, and Grass Valley and Nevada City, in Nevada County. This service was essential to both counties, as were the depot stops en route. The trains transported both freight and passengers. The monument was moved to Grass Valley Street in 1914 and torn down by 1920. (CHM.)

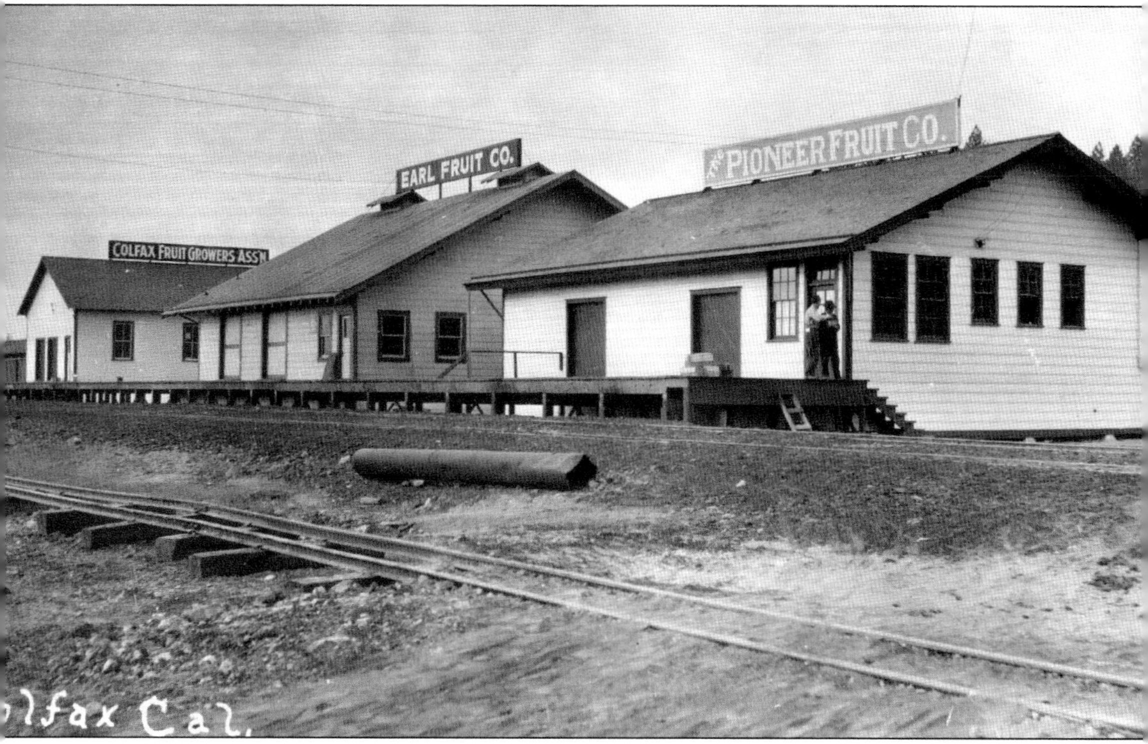

Colfax and its neighboring communities were renowned for the quality fruit they grew. Three fruit-packing companies operated facilities along the train tracks: Earl Fruit Company, Pioneer Fruit Company, and the Colfax Fruit Growers Association, formed in 1917. Apples, pears, and peaches grown in the foothill orchards were touted as being of higher quality than those from the valley. Once the railroad offered refrigeration, farmers had the capability to ship their produce all over the world. Icing stations were built at regular intervals along the tracks. There, new blocks of ice would replace melted ones. Men lowered each 300-pound block through a hatch and made sure the car was full and tightly packed. Ice was harvested from lakes and frozen ponds by sawing it into large blocks. By the 1950s, the railroads chilled the cars mechanically. (AS.)

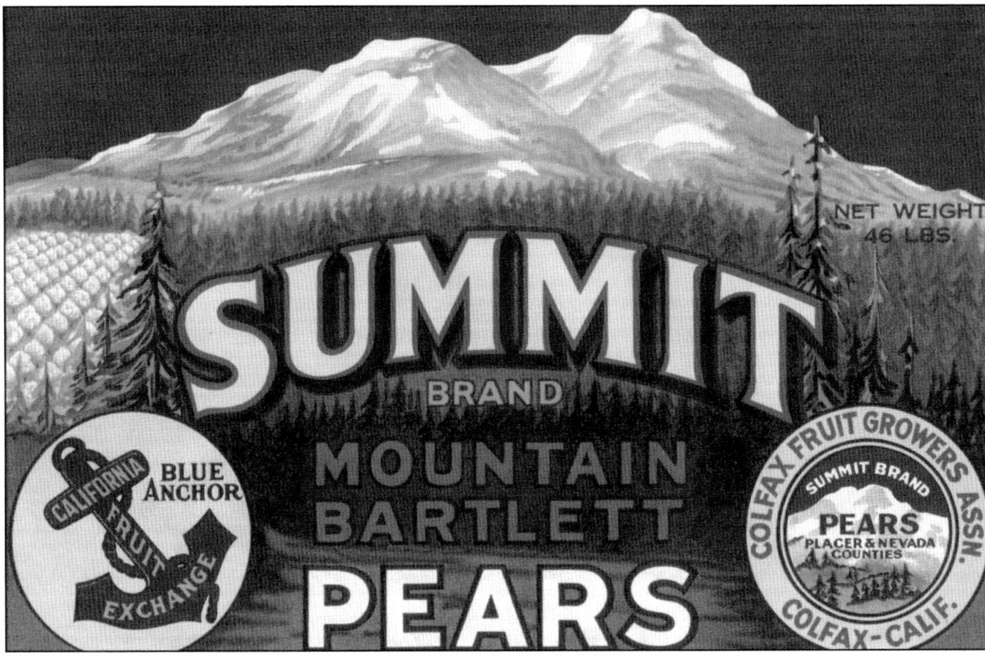

The designers for fruit boxes adjusted them to minimize spoilage from the constant motion of the trains as they traveled across the country. The labels boasted of the origins and perfect nature of the products. Bright, bold color combinations complemented the artistic designs ornamenting the crates. These two pear box labels tell the brand, weight, and packing company association. The top label shows the idyllic scenery of the Sierra Nevadas surrounded by forests of evergreens and highlighted with orchards on the left. These special labels have become collector's items and are displayed in train museums like the California Train Museum in Sacramento. Look for the telltale blue anchor of the California Fruit Exchange on Sun Smile, Mountain Bartlett, and Colfax Fruit Exchange labels. (Author.)

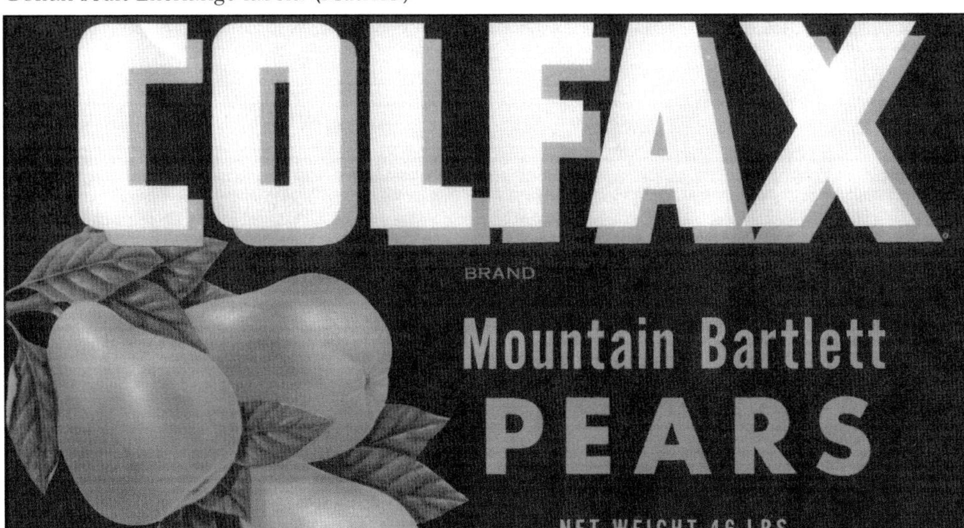

In the 1880s, peaches and apples dominated the foothill farms. Over the next 50 years, Bartlett pears, Tokay grapes, and Hungarian prunes took the lead. Bartlett pears grew especially well in the foothills and were in high demand. (Author.)

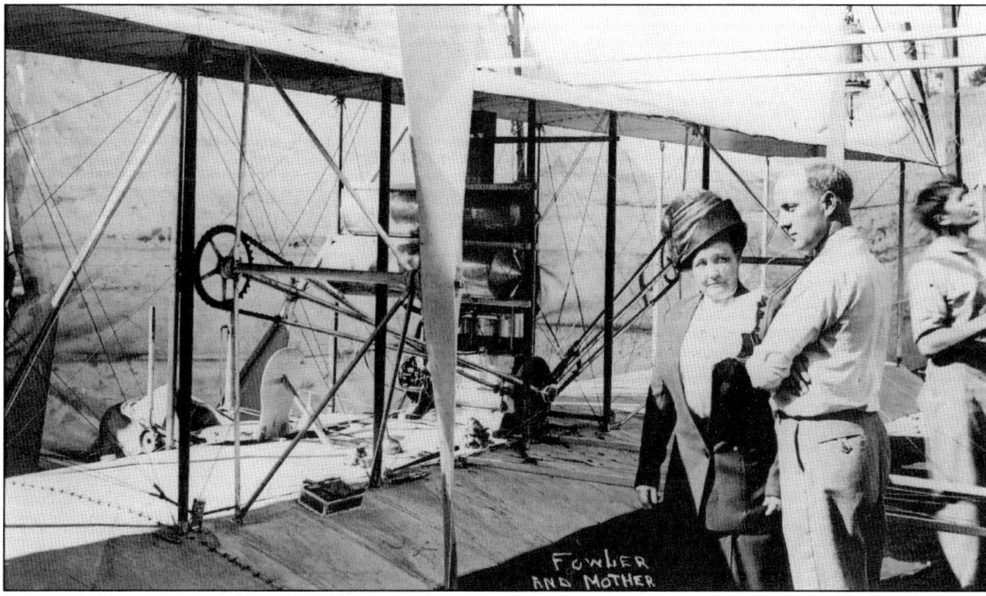

Railroads were not the only form of transportation available in the new century. Inventors worked to improve airplanes so they could fly over high mountain ranges. Robert G. Fowler attempted to be the first pilot to tackle the Sierra Nevadas. Fowler is seen here with his mother, Mary Frances "Faye" Fowler, after the flight had stopped in Auburn and then crashed at Lake Alta above Colfax. He was not injured in the accident. In the photograph, they await repairs to the plane in Colfax. (PCM.)

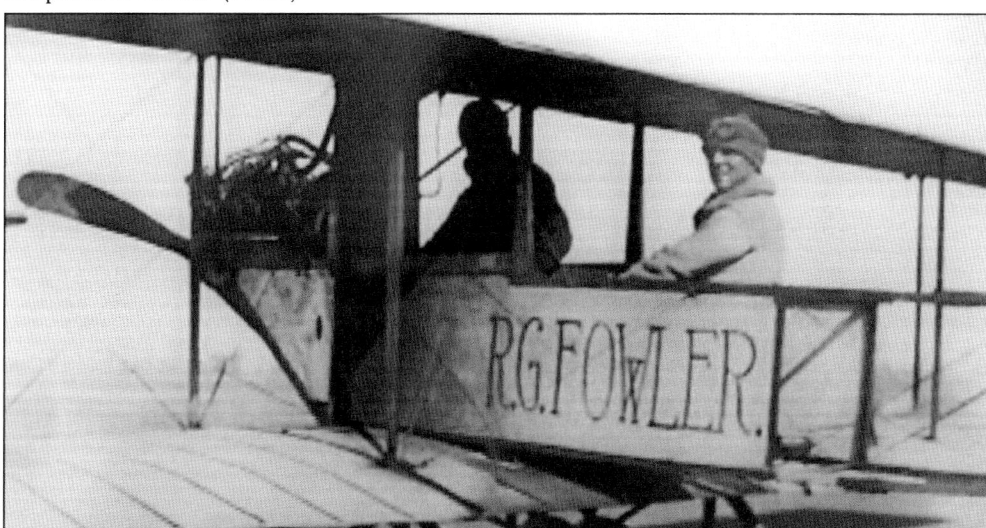

Fowler is seen here in his plane. There is a 1997 plaque dedicated to him in the town of Colfax near the railroad tracks, where the actual Model B biplane was rebuilt after it crashed in Alta. William Randolph Hearst had offered $50,000 to the first pilot who crossed the continent by October 11, 1911. Fowler became determined to win the prize and began his flight on September 11, 1911, in San Francisco, California. After landing in Sacramento, he continued on up to the Sierras. A radiator problem forced him down when the engine overheated, and one wheel was damaged on the landing. After repairs in Colfax with parts from Taylor's Foundry in Grass Valley, Fowler tried again from Southern California. Unfortunately, no one received the coveted prize. (CHM.)

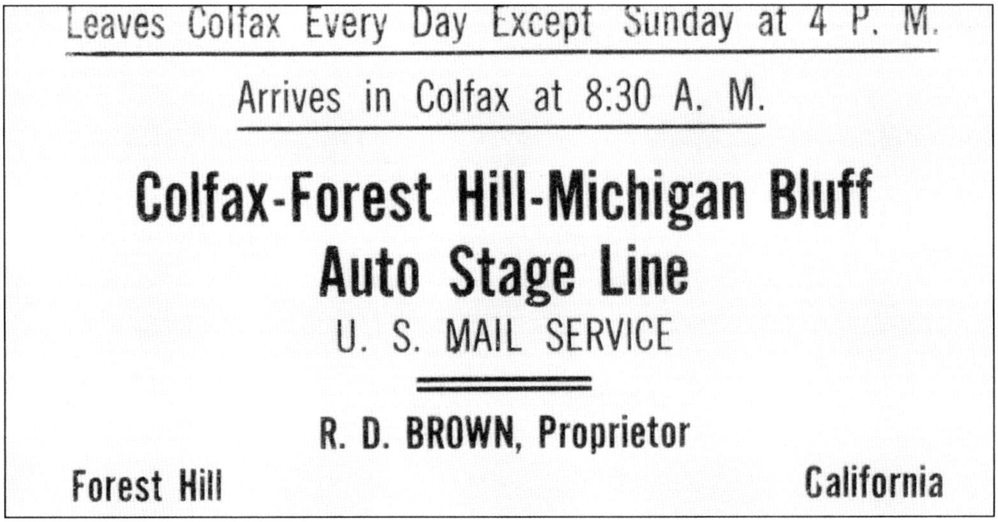

Leaves Colfax Every Day Except Sunday at 4 P. M.

Arrives in Colfax at 8:30 A. M.

Colfax-Forest Hill-Michigan Bluff
Auto Stage Line

U. S. MAIL SERVICE

R. D. BROWN, Proprietor

Forest Hill **California**

An auto stage line was put into service with rounds between Colfax, Forest Hill, and Michigan Bluff. It provided mail delivery service for the area and operated every day except Sundays. This card represents either a ticket or an advertisement for the service and is thought to be from about 1915, according to the proprietor's great-grandson R.D. Brown. (PCM.)

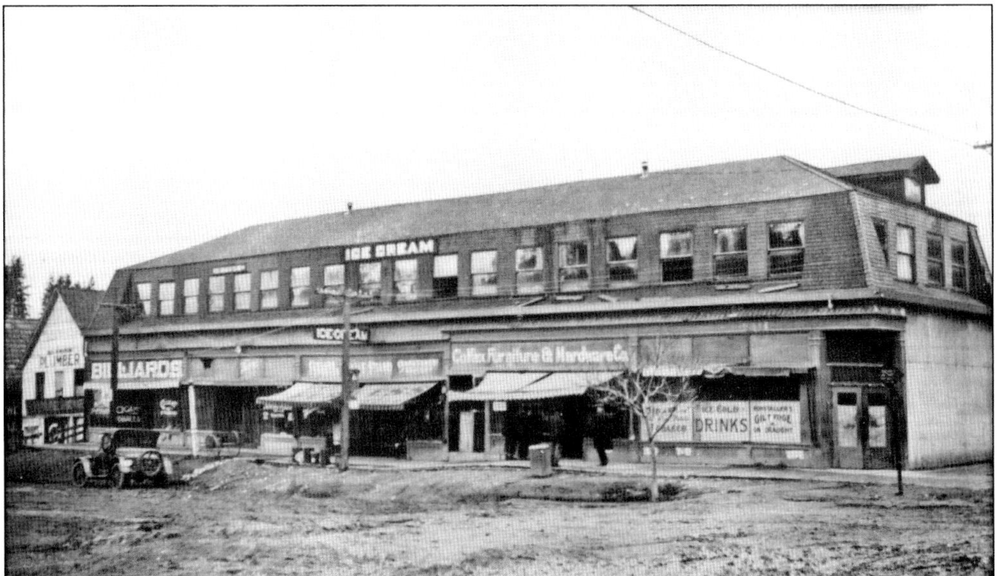

The Lobner building contained numerous businesses, from furniture and grocery purveyors to billiards halls and ice cream shops. The Colfax Furniture and Hardware store sat in the middle. Next to the Lobner building was a much smaller building housing a plumbing business. The car on the left sets this photograph in the 1920s. The Lobner building burned in 1932. (AS.)

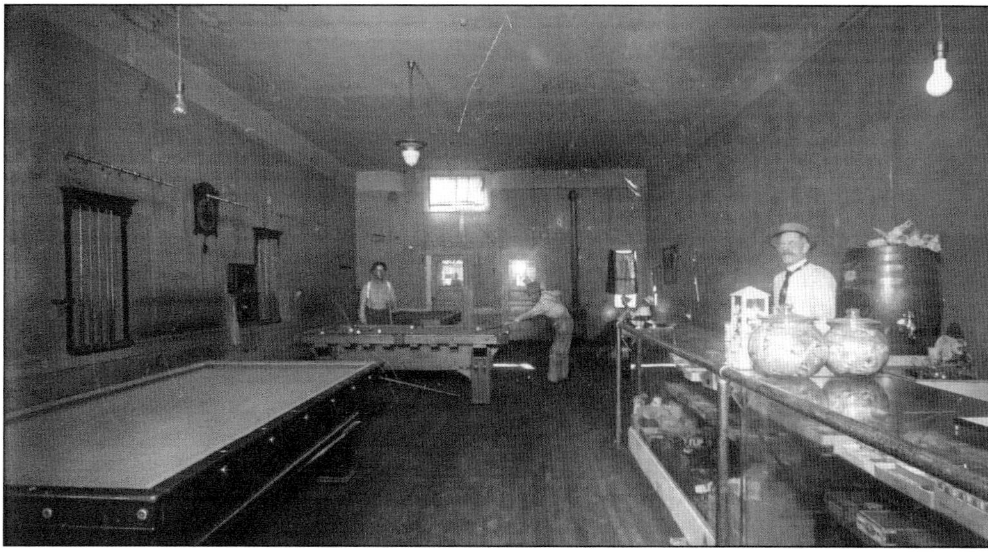

This pool hall was most likely operated by Fred Schultz. Cigars, the fashionable smoke of the day, were for sale in the glass case. (CHM.)

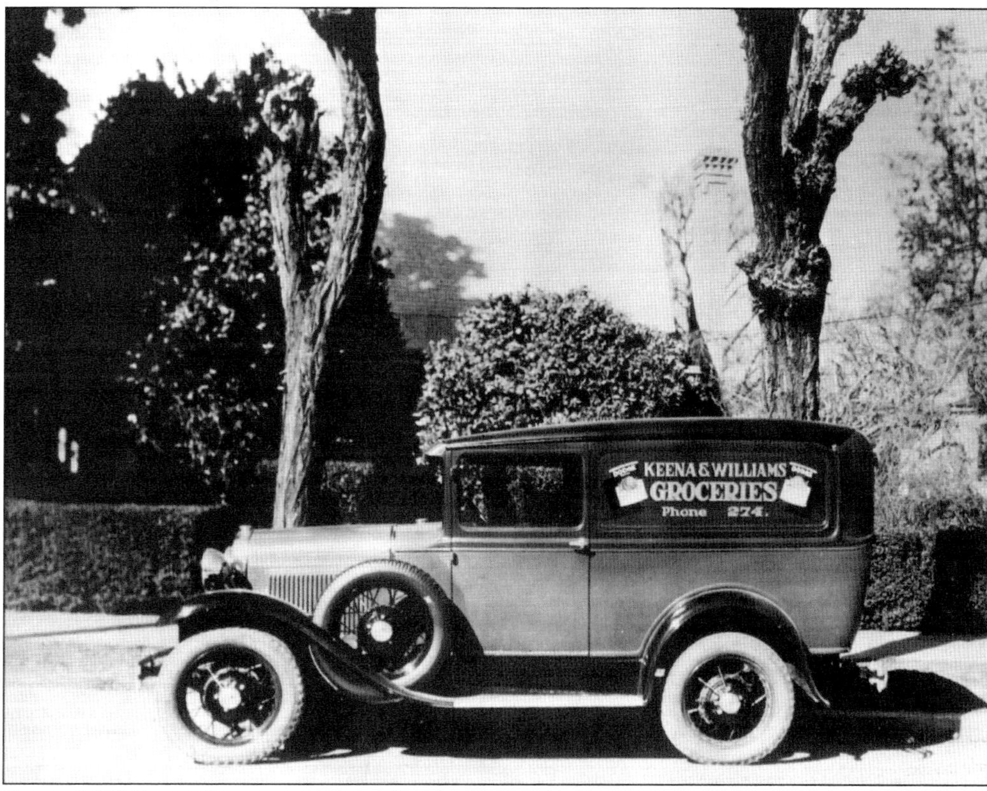

Keena & Williams Groceries delivered orders to local customers in its classy Dodge delivery van. Charles Keena and Albert Williams began their partnership in 1914. According to *History of Placer County*, they were "among the most prosperous businessmen in Placer and Nevada County." In 1930, telephone numbers contained only three digits, as seen here, and calls were still connected by an operator. Keena married Lorraine Lobner. (CHM.)

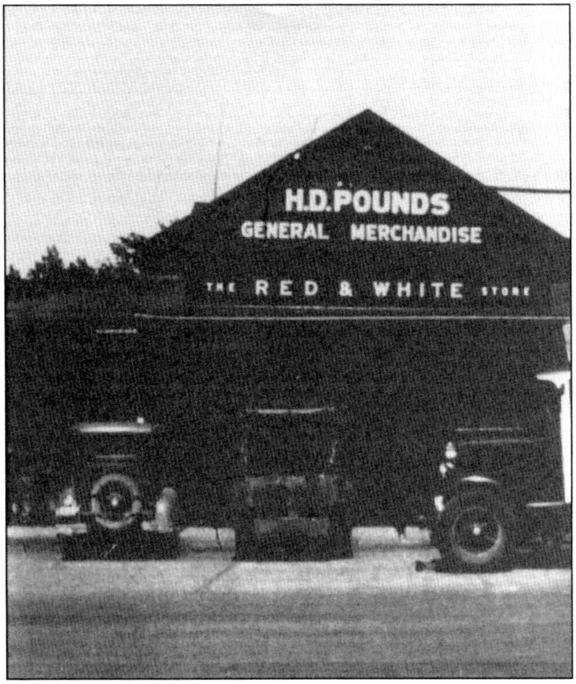

H.D. Pounds owned a general merchandise store at the corner of Grass Valley and Main Streets. He bought the store in 1922 with Walter Viscia. Later, Pounds bought out Viscia. On December 27, 1931, the store's merchandise burned and the building was bought by the Purity Store. There is still a store at that location. (CHM.)

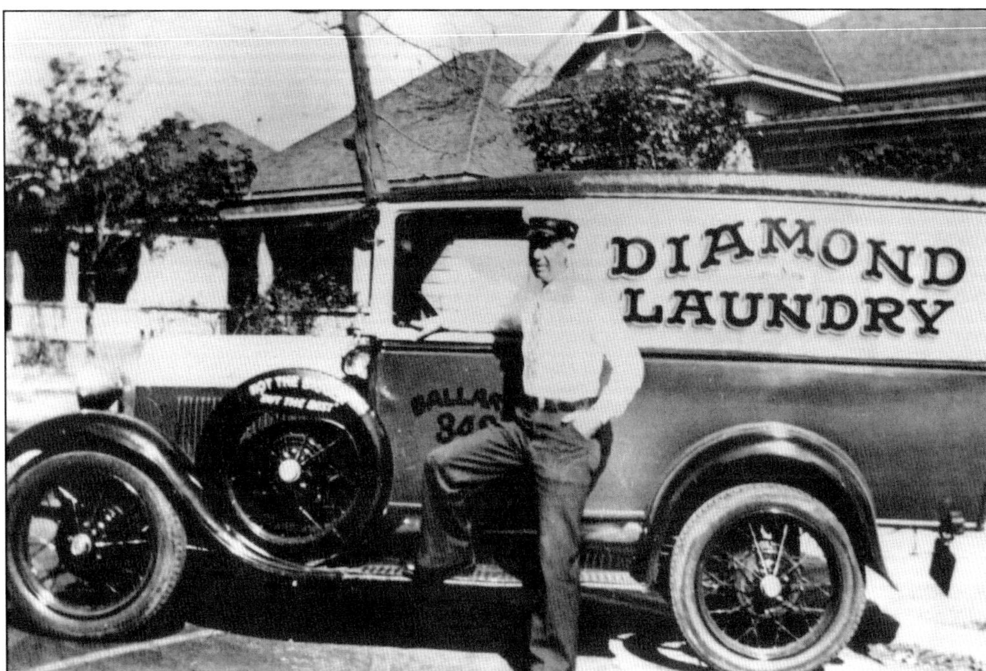

This deliveryman stands with his foot on the running board, posing for a commercial shot. The delivery van is a 1930 Ford Model A. Diamond Laundry delivered laundry to its customers, who likely included businesses as well as individuals. The Cortespassi family founded a steam laundry in town in September 1922, although it is unclear if this is it. This laundry location was originally occupied by the Colfax yard of the Auburn Lumber Company, owned by S.K. Williams. The True Value hardware store and Colfax Lumber eventually took over the site. (CHM.)

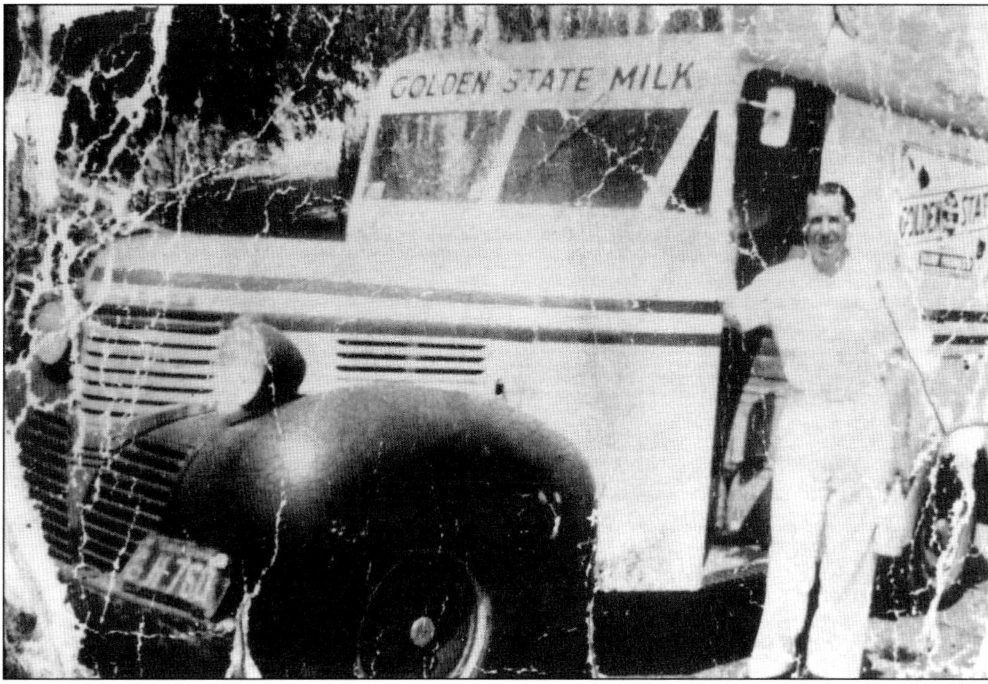

Colfax local Walter Keck is seen here in front of the 1939 Dodge Golden State Milk delivery van he drove. Home milk delivery was common in many towns at this time. Golden State Milk used the square Blake-Hart milk bottles, which are now treasured antiques. Blake had a dairy in Sacramento, and Blake and Hart patented the unique bottles in 1927. (CHM.)

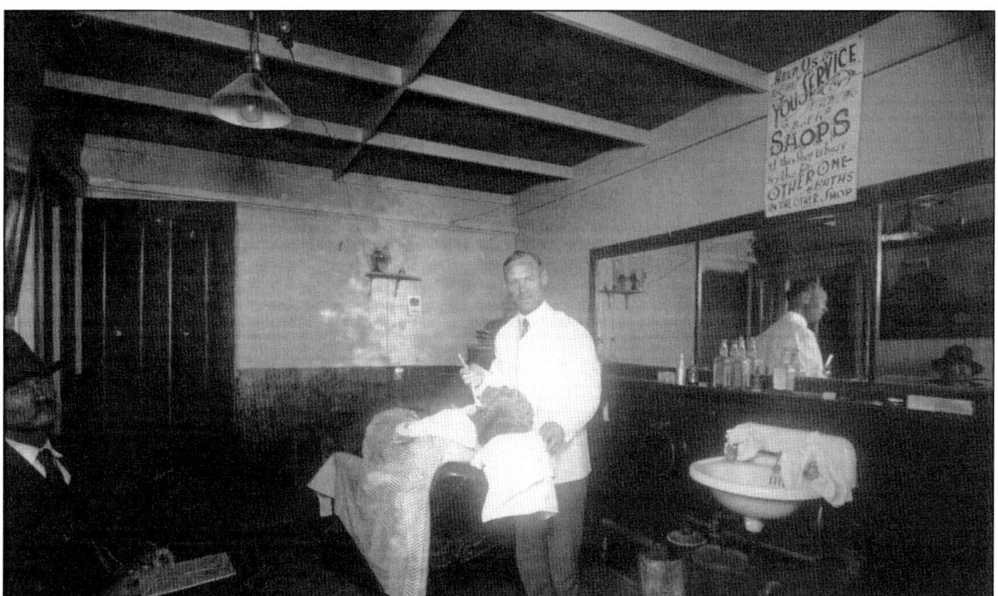

Dave Oswald Marson busily shaves a man with a straight razor in this post–World War I image. A shoemaker plied his trade next door, and Marson's uncle Oswald Marson built an L-shaped building with a haberdashery behind where the shoemaker had been. (CHM.)

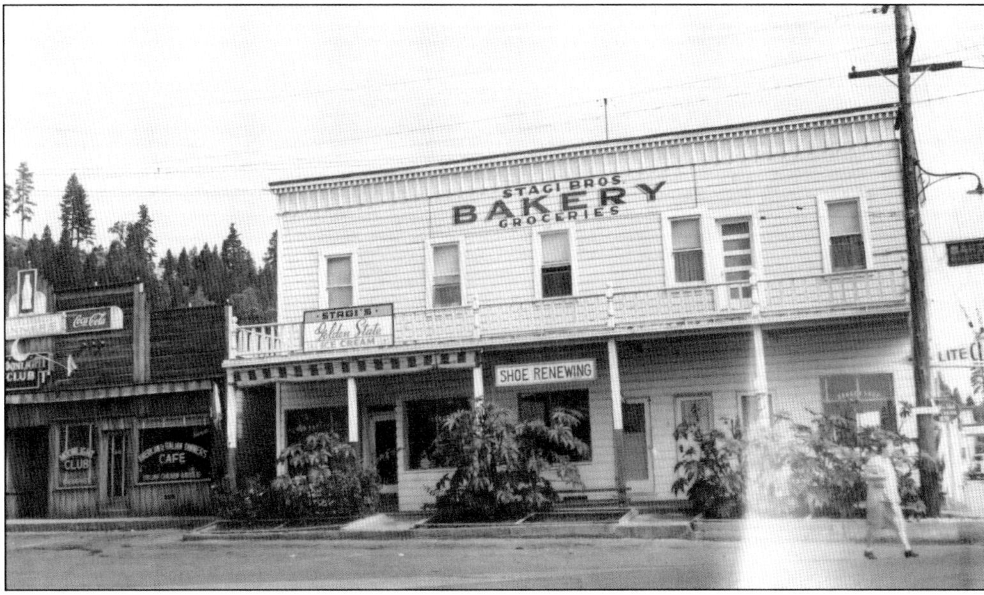

Stagi Bros. Bakery and Groceries first opened in 1907 on Depot Street at the end of Main Street. The Stagi brothers proudly carried Golden State Ice Cream for their customers. A shoe repair shop also occupied the building. Later, Frederick Pitz operated the bakery for 20 years. Café Luna occupies the space today. (AS.)

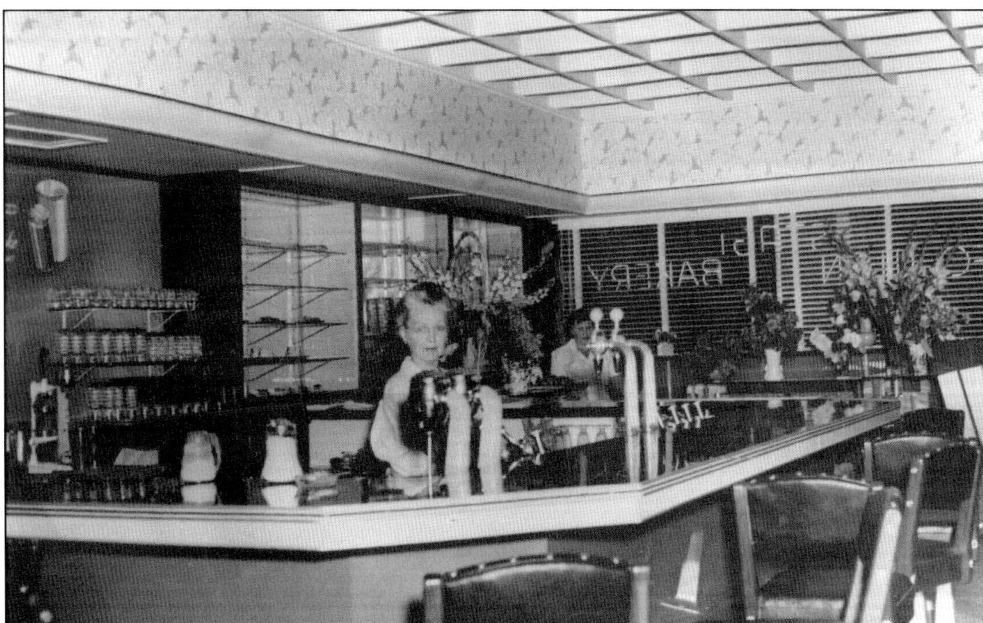

This interior view of the bakery shows Lil Pitz working at the soda fountain and the fashionable leatherette stools for customers to relax in and chat over an ice cream sundae or a soda. An old-fashioned milk shake machine is tucked along the counter by the glasses, close to the sugar and creamers. (CHM.)

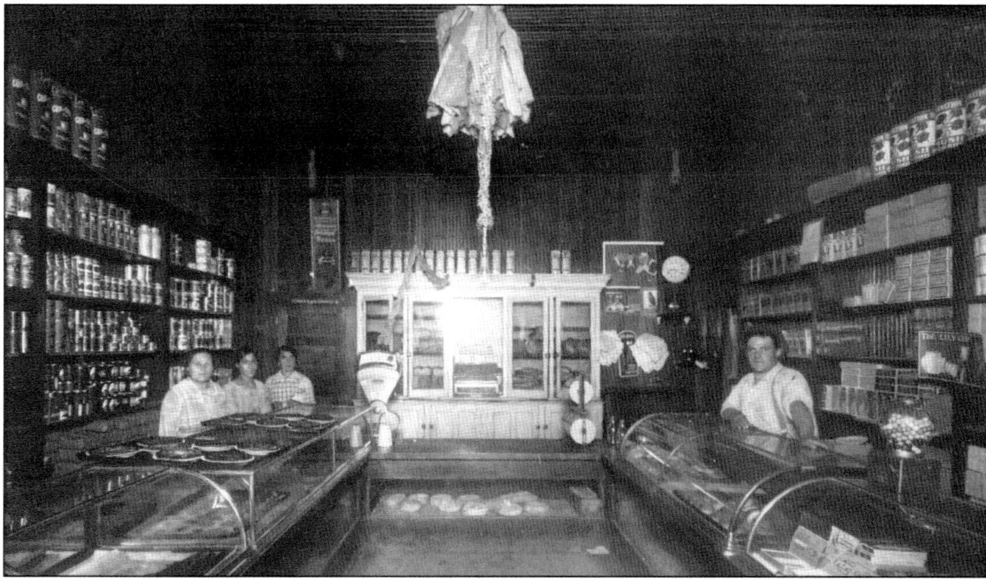

Above, in the 1920s, the Corner Market appears to have offered everything from canned goods to baked goods like bread and fresh pies. Even cigars awaited purchase in their own glass display case. The market stood at Main and Grass Valley Streets. In 1870, Hayford, Hines & Company established a general merchandise store there. By 1879, it became Hayford, Perkins & Company. In 1922, Walter Viscia moved his store there from its previous location at Front and Main Streets. (CHM.)

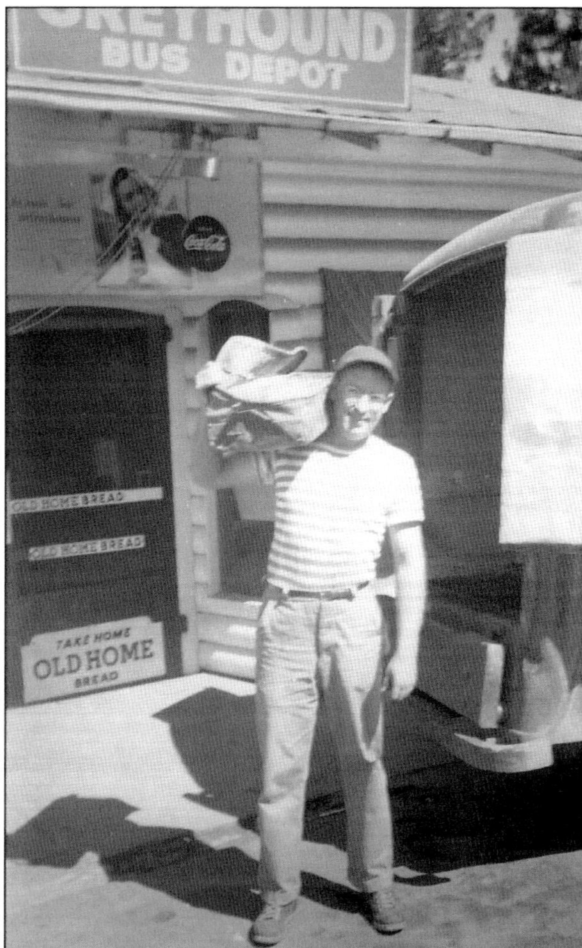

This gentleman carries what appears to be a burlap sack containing a block of ice into a bakery in the Colfax bus depot. The man is likely Carl Bell, the owner and agent for the Union Ice Company and Kenison Soda. Bell also owned a bottling business and manufactured soda and soft drinks. The bus depot sign and the delivery truck model date the photograph to the 1930s. The ice plant was where Performance Marine now stands on North Main Street, just up from School Street. (JLJ.)

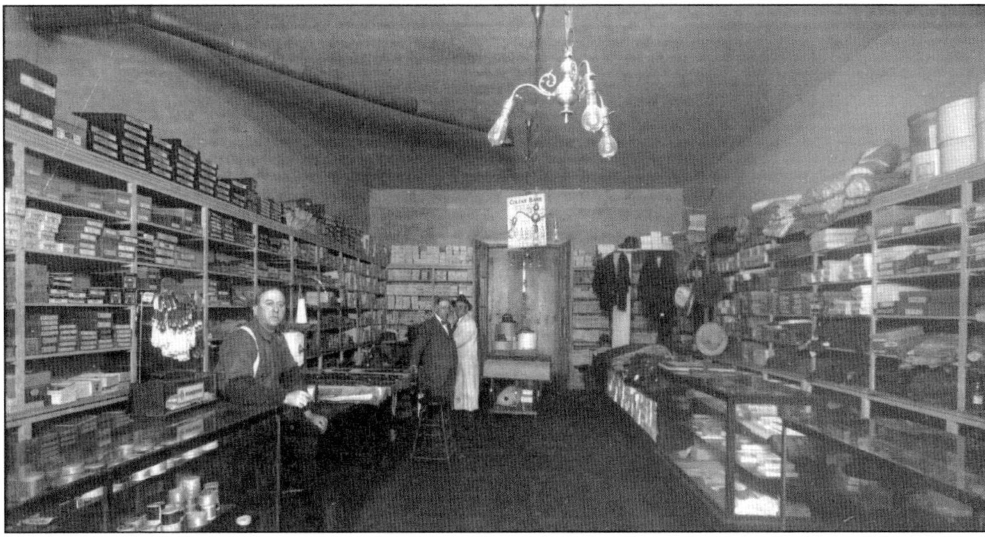

This photograph peeks inside Marson's Men's Store in the 1920s. It appears to be a typical haberdashery of the era, carrying everything necessary to dress a fine gentleman, including hats—in the glass case on the lower right—clothing, handkerchiefs, and shoes. (CHM.)

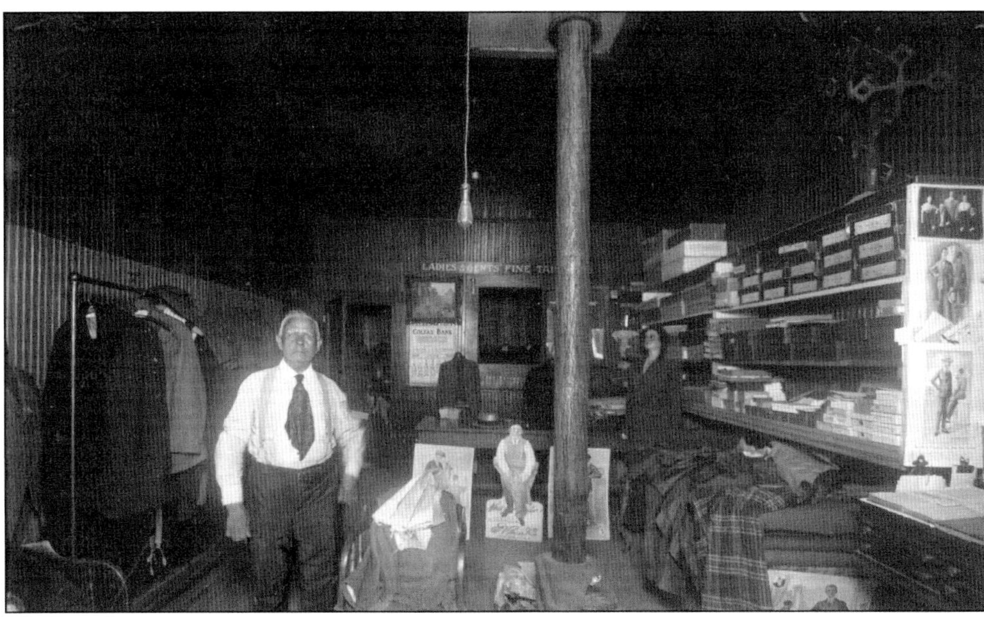

This store, seen here in the 1920s, is believed to be Petri's store, with a sign for "Ladies & Gents' Fine Tailoring" above the counter. Many bolts of suit fabric lined the right side, with a few pulled out for viewing. At the time, the men's and women's tailoring sections would have been separated for propriety. (CHM.)

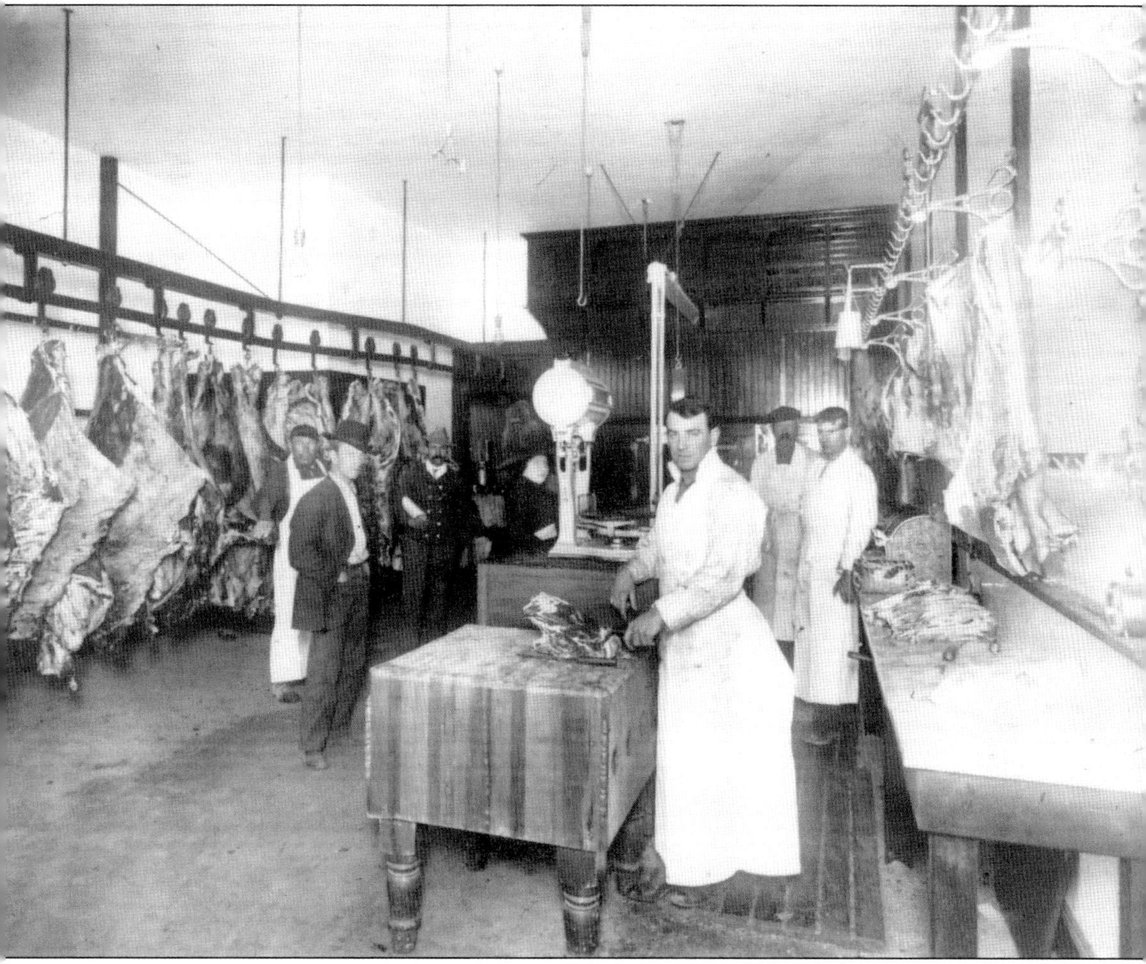

This graphic portrayal titled *Melvin Russell in Butcher Shop* shows butchers at work around 1910. Three men with blood-smeared aprons take the sides of beef hanging on hooks and carve them into the cuts their customers have ordered. According to the historic plaque, in 1875, John Mischler owned the building, which was built by Daniel A. Russell. The next year, the NCNGRR train struck Mischler as he walked home. The conductor warned him with a whistle and slowed down the train, but the cattle guard snagged Mischler. His brother, Martin Mischler, operated the shop for a year then sold it. It changed hands multiple times over its history, but when J.L. Stokes, a member of the Russell family, took it over in 1894, it remained within his family for 83 years, until 1977. (PCM.)

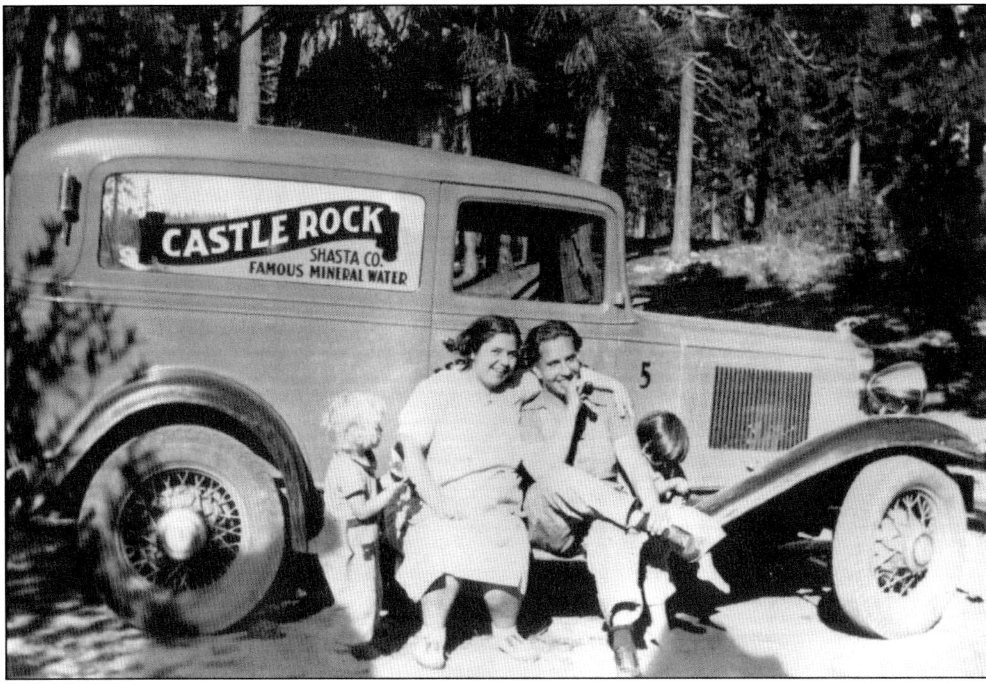

The Castle Rock Shasta Co. delivered its "famous mineral water" to customers in Colfax. Trains transported the water in special cars from Dunsmuir, California, where the water originated in the lava tubes and streams at the base of Mount Shasta. Castle Rock appears to have been a family endeavor, as the young couple sits on the running board taking a break in 1930. (CHM.)

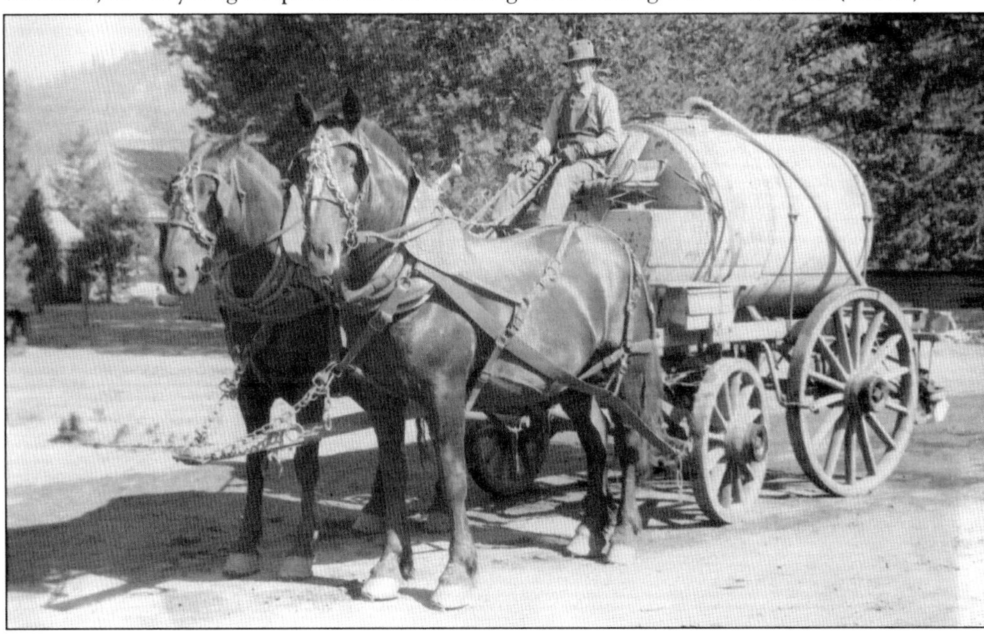

Water wagons delivered water around the area for commercial and residential use to those who did not have their own wells. Hanson and Wills owned the water delivery business, which was known for its horse-drawn wagon equipped with the huge tank. This photograph was taken in 1913. (CHM.)

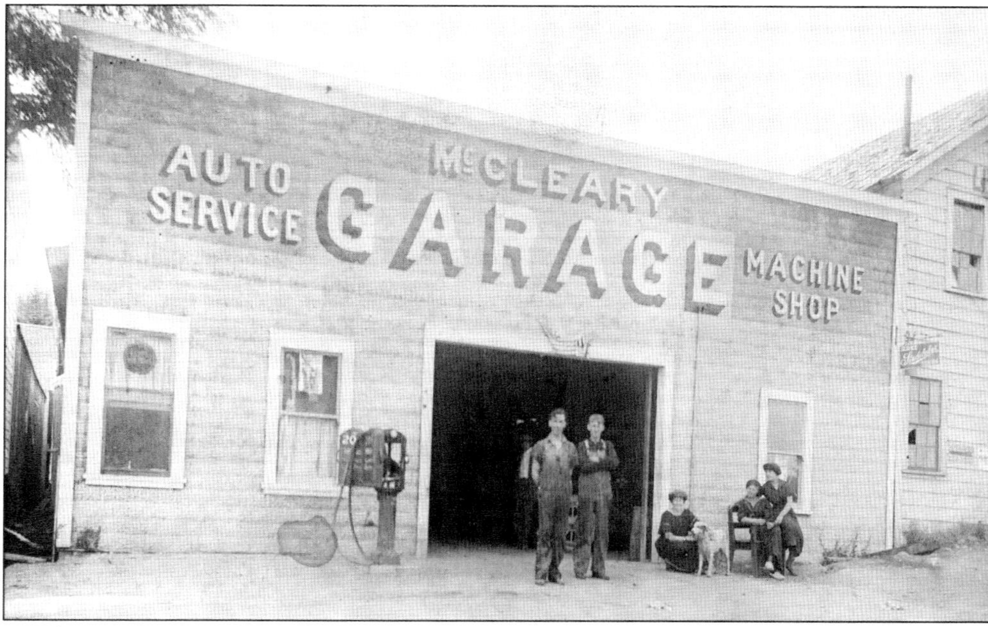

McCleary Garage, with its Standard Oil gas pump out front, serviced automobiles and operated a machine shop. The garage, owned by John McCleary, was at the corner of School and Main Streets. Originally trained as a blacksmith, he switched to working on automobiles when they became popular. He also sold Ford and Dodge models. McCleary arrived in Colfax in 1872 and opened his livery. His son, John B. McCleary, ran the garage and machine shop after him and later became an agent for Packard and Ford automobiles. (CHM.)

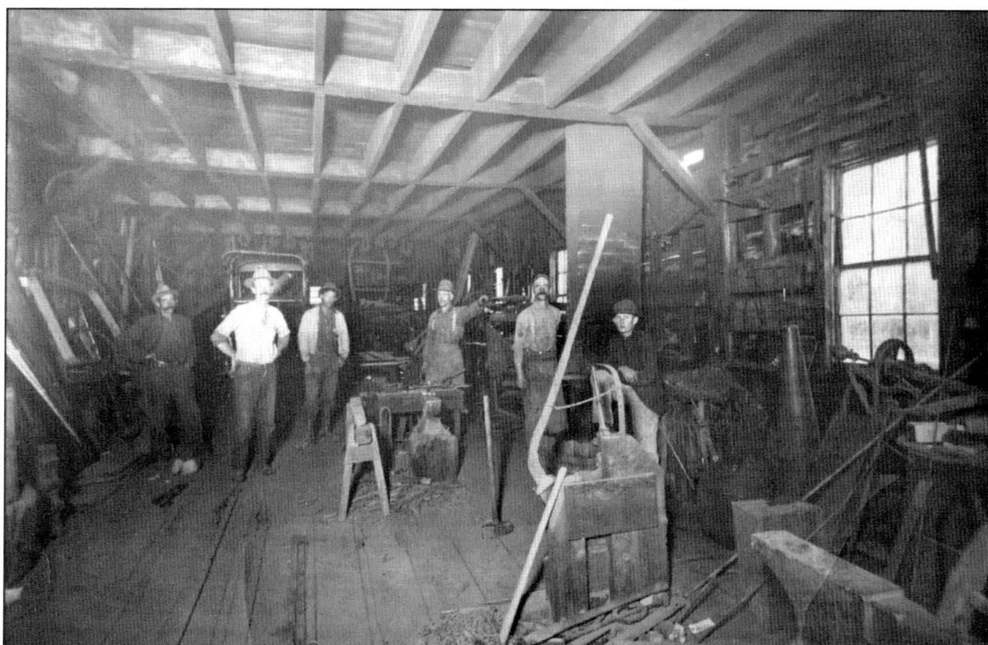

These men pose for a photograph in McCleary Garage. Owner John McCleary is in the white shirt with his hands on his hips. (CHM.)

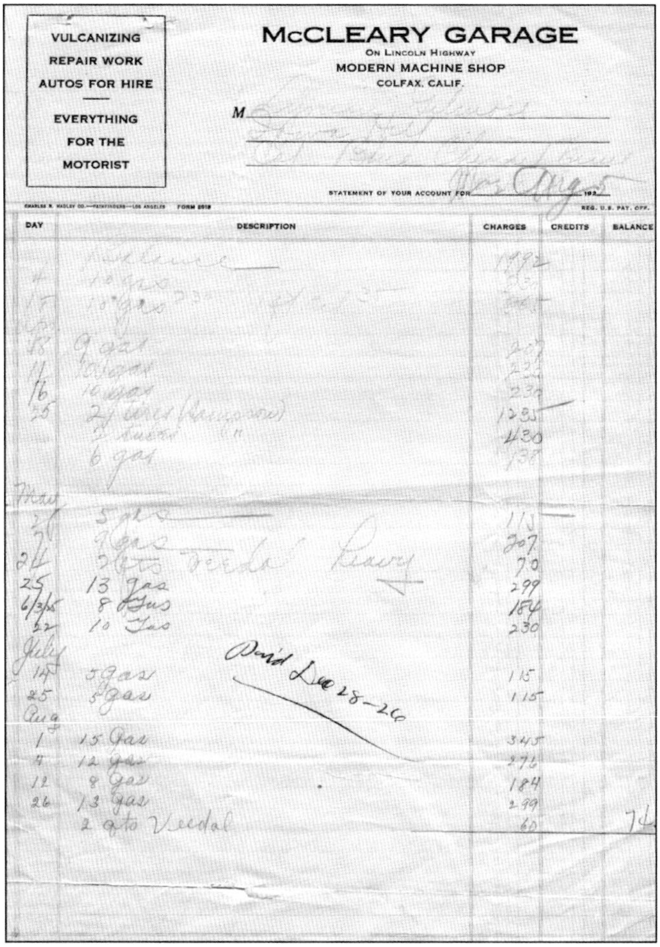

VULCANIZING
REPAIR WORK
AUTOS FOR HIRE
EVERYTHING
FOR THE
MOTORIST

McCLEARY GARAGE
ON LINCOLN HIGHWAY
MODERN MACHINE SHOP
COLFAX, CALIF.

STATEMENT OF YOUR ACCOUNT FOR

DAY	DESCRIPTION	CHARGES	CREDITS	BALANCE
25	13 gas	299		
6/3/25	8 Gas	184		
22	10 Gas	230		
July 14	5 gas	115		
25	5 gas	115		
Aug 1	15 Gas	345		
12	8 Gas	184		
26	13 gas	299		
	2 qts Veedol	60		

Paid Dec 28-26

This bill issued by McCleary Garage indicates that Lyman Gilmore, an inventor and aviation pioneer, paid it on December 28, 1926. The garage charged 23¢ per gallon for gasoline and 70¢ for two quarts of heavy oil. (AS.)

Below, Ford Model Ts are lined up for servicing inside McCleary Garage as the crew looks on. (CHM.)

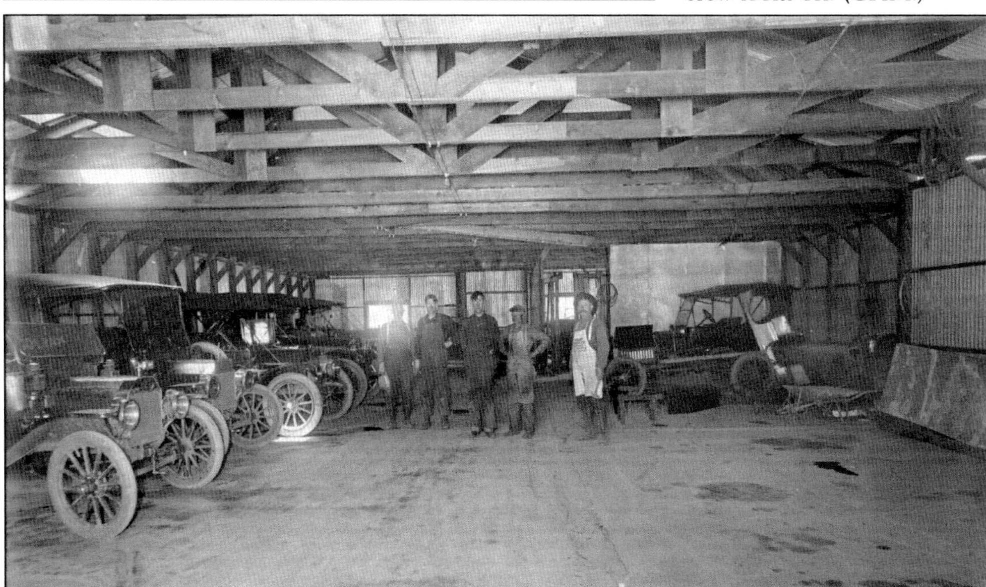

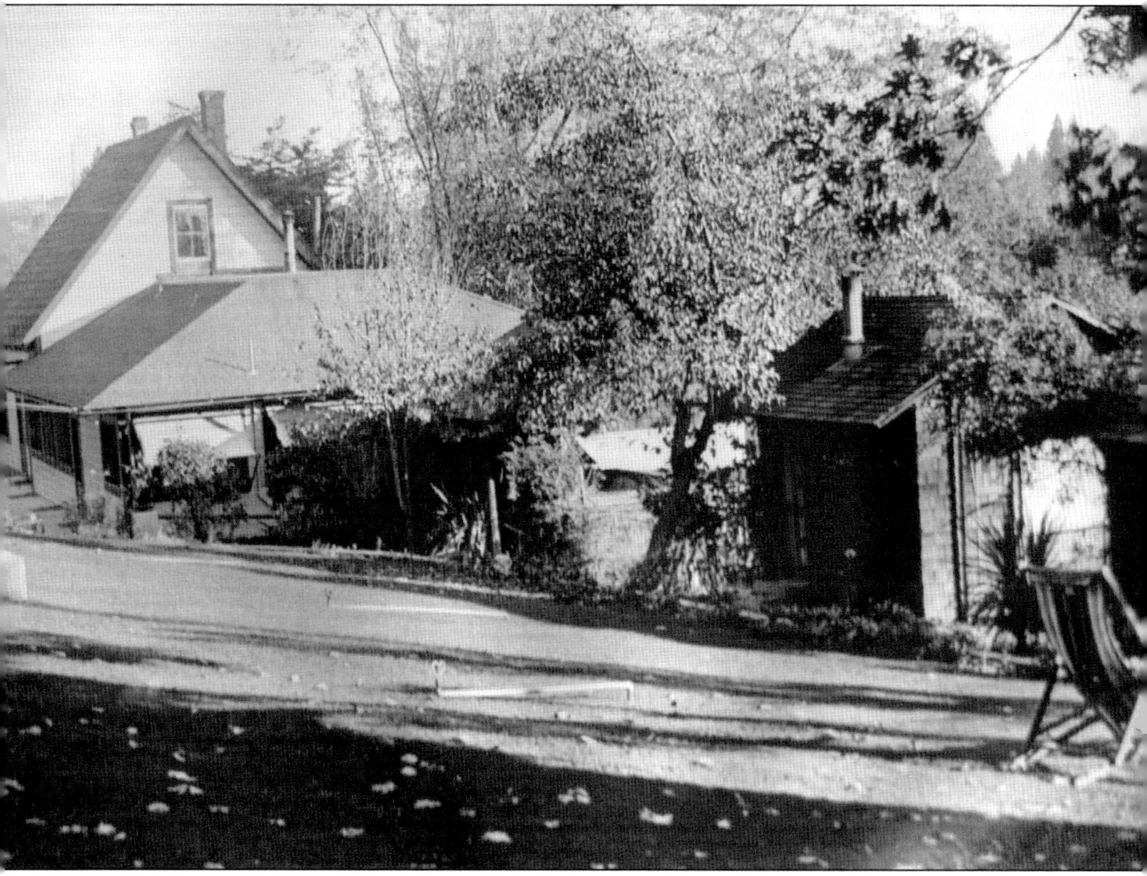

Colfax became famous for its care of tuberculosis patients. Dr. Robert A. Peers is credited with assessing the area for suitability when he moved west in 1899. He found the climate and the elevation—2,400 feet to 2,500 feet—ideal for treating the disease. The Colfax Hospital and cabins were set up on Kneeland Street in town. Dr. Peers, Dr. Padgett, Dr. Dunard, Mrs. Weaver, and Mrs. Potter worked there. Peers began the Colfax School for Tuberculosis with just a few beds but grew it to 180 beds. Several sanatoriums were established in the foothills under Dr. Peers's supervision. The five facilities represented the largest group of private tuberculosis treatment centers in the United States. (CHM.)

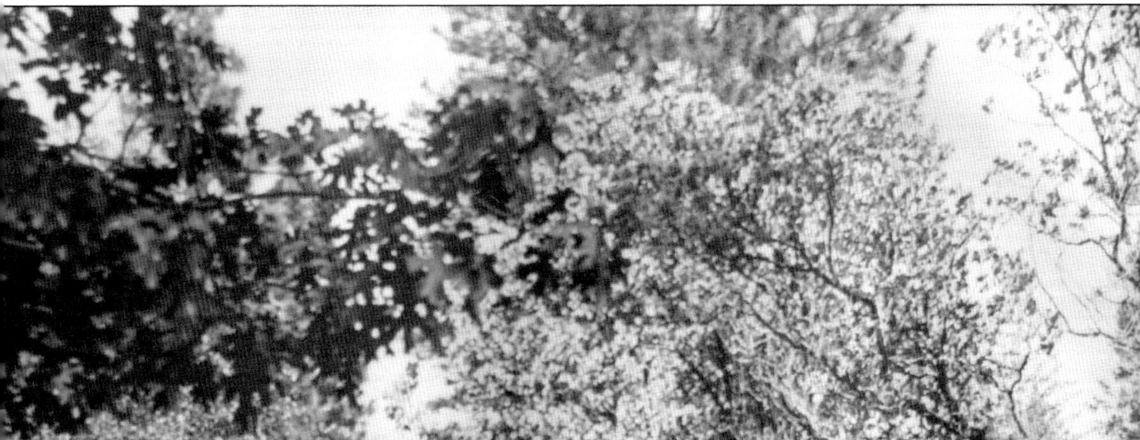

The Colony Company built another tuberculosis treatment facility with cabins for patients on Colony Road off of Tokayana Road in Colfax. Standard Oil also had its own complex for employees afflicted with the disease. Sanatoriums remained active until the incidence of the disease abated dramatically with the introduction of antibiotics. The cabin windows in sanatoriums did not have panes, to allow the constant flow of fresh air. Large porcelain bottles filled with hot water warmed the occupants' beds at night during the cold weather. (CHM.)

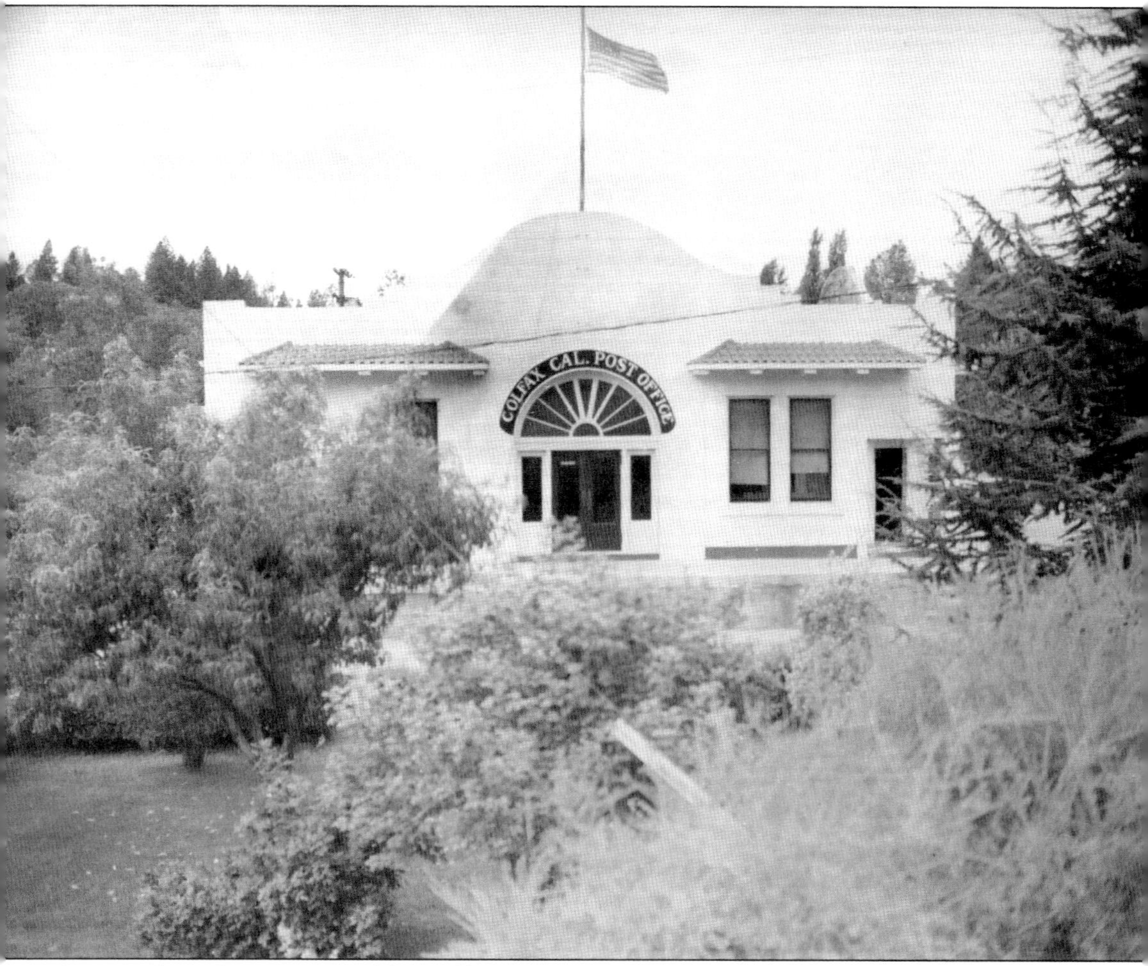

Starting the 1850s, water was delivered by canals and ditches. Once agriculture took hold in the area, irrigation water was in high demand. In the 1890s, a new distribution system, with flumes spanning up to 60 miles, was built to meet this need. In 1902, Central California Electric Company established the first hydroelectric company through its subsidiary, South Yuba Water Company. Pacific Gas and Electric (PG&E) took over South Yuba Water Company in 1905. After the new post office building was constructed in Colfax in 1917 at the corner of Main and Church Streets, where the library is today, PG&E relocated from its previous location in the Lobner Building, and it shared this site with the post office. (AS.)

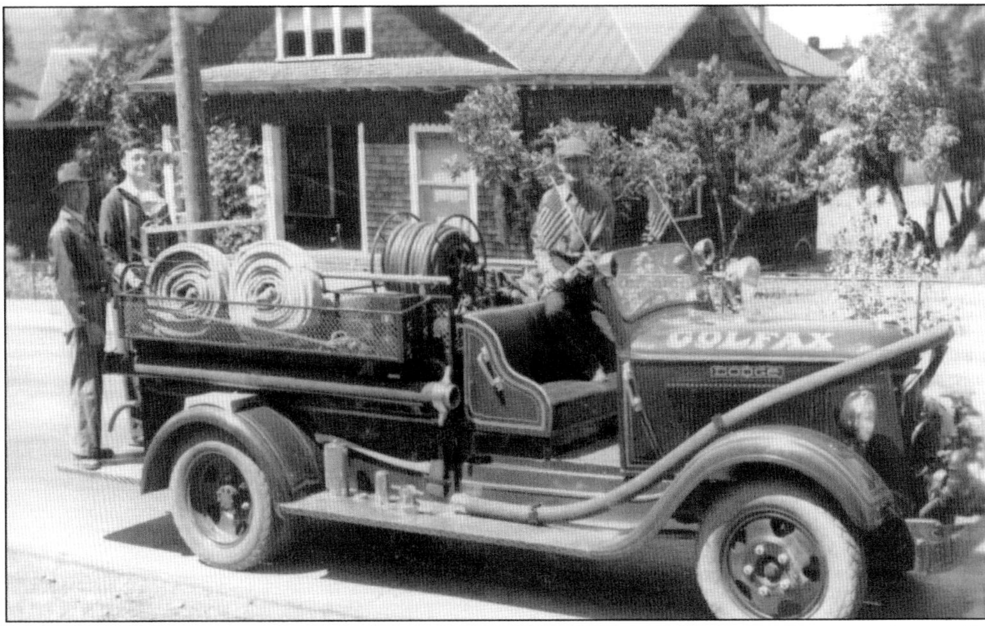

Volunteers fought fires that erupted in town and maintained all the equipment. Originally, citizens manned bucket brigades that proved ineffective in controlling the frequent fires that ravaged historic wooden towns. The Colfax Volunteer Fire Department (CVFD) was formally organized in April 1910, although the volunteers continued to operate with just bucket brigades and hose-wagons until the department purchased its first fire truck in 1923. Seen here are, from left to right, Mr. Frederickson, the constable of Colfax; Francis West; and Chuck Martin, the driver. CVFD is now managed by CAL FIRE. (AS.)

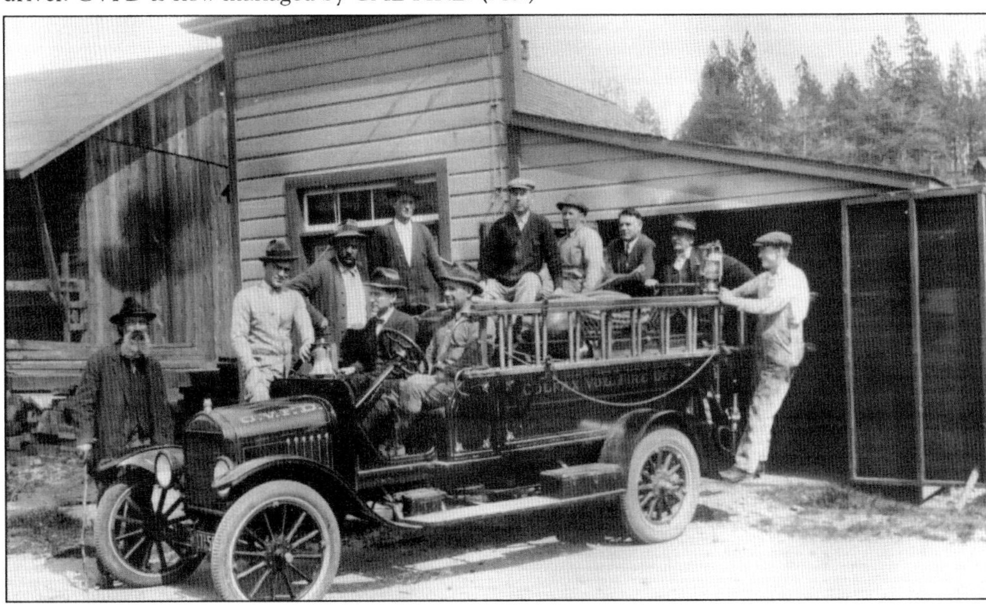

Seen here in front of the new CVFD quarters around the new ladder truck are volunteers, from left to right, Ed Atwell (standing), L. R. Starr, Jack Viscia Sr., N.E. Booth, F.D. Kuenzly, Ben Dressen, S.H. Sayre, Walter Viscia, and John B. McCleary; (seated) H.D. Pound (assistant chief) and C.P. Wills (chief). (CHM.)

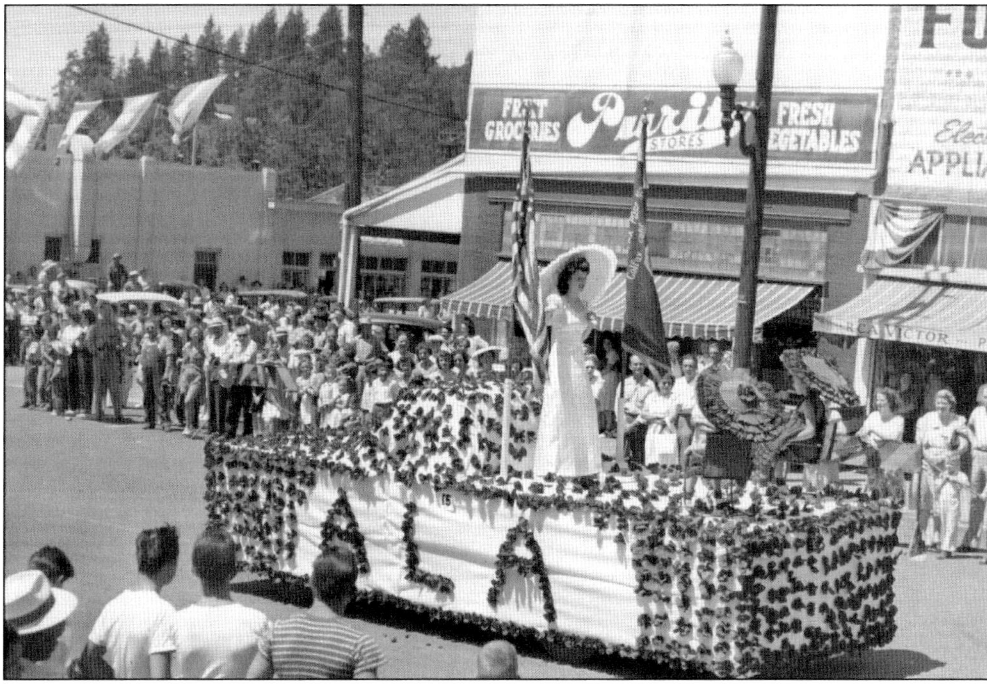

Like most small towns, Colfax loved any excuse to gather everyone together and celebrate. These two photographs show elaborate floats in a 1936 parade, with townspeople packing both sides of Main Street. Here, the float drifts past the Purity Store and the appliance store. (CHM.)

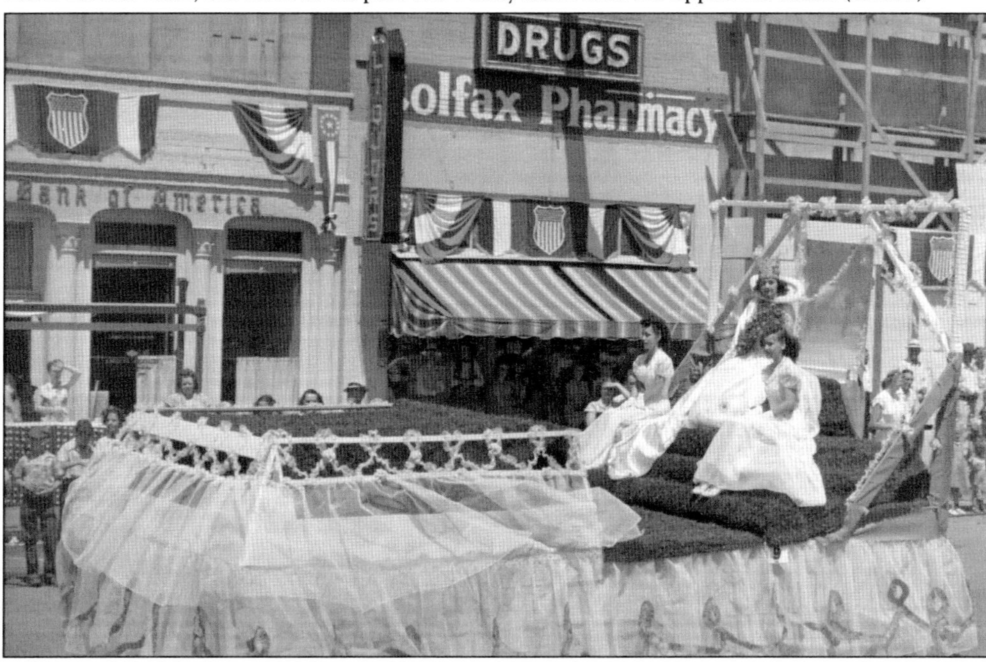

This view shows the parade passing by Bank of America and Colfax Pharmacy. The Soroptimists' royal float, seen here, is covered with flounce and formality. The Colfax queen, surrounded by her attendants, proudly waves at the crowd. The Soroptimists produced a special float for the parade each year. (CHM.)

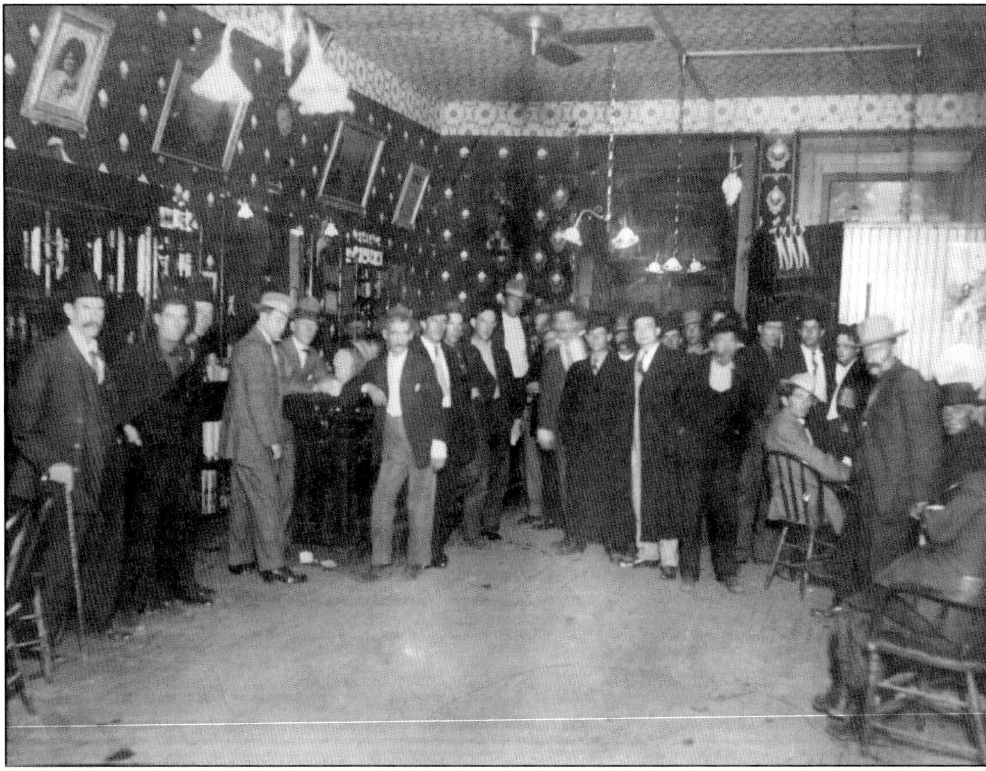

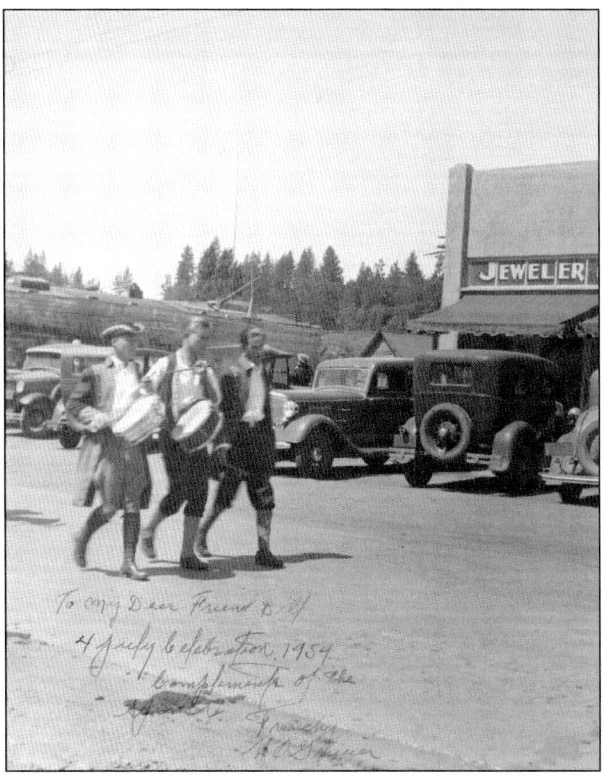

This 1920s photograph shows an elegant bar in Colfax. Though not identified, it appears to be the inside of a hotel, judging from the Victorian wallpaper, decorative light fixtures, and paintings. Many finely dressed gentlemen posed for the photograph. (CHM.)

The jewelry store in the background belonged to H.O. "Frenchy" Grenier. After the Fourth of July celebration in 1934, he autographed this parade photograph of the drummers marching. Grenier signed the memento, "Complements of the Sponsor." (CHM.)

H.O. "Frenchy" Grenier poses on Main Street in Colfax in 1938. Note the change in style of the cars from the previous photograph, which was full of Ford Model As. He is decked out in his riding crop and tall boots. Grenier was born in 1880 and died in 1953 in Colfax. He is buried in the local cemetery. (CHM.)

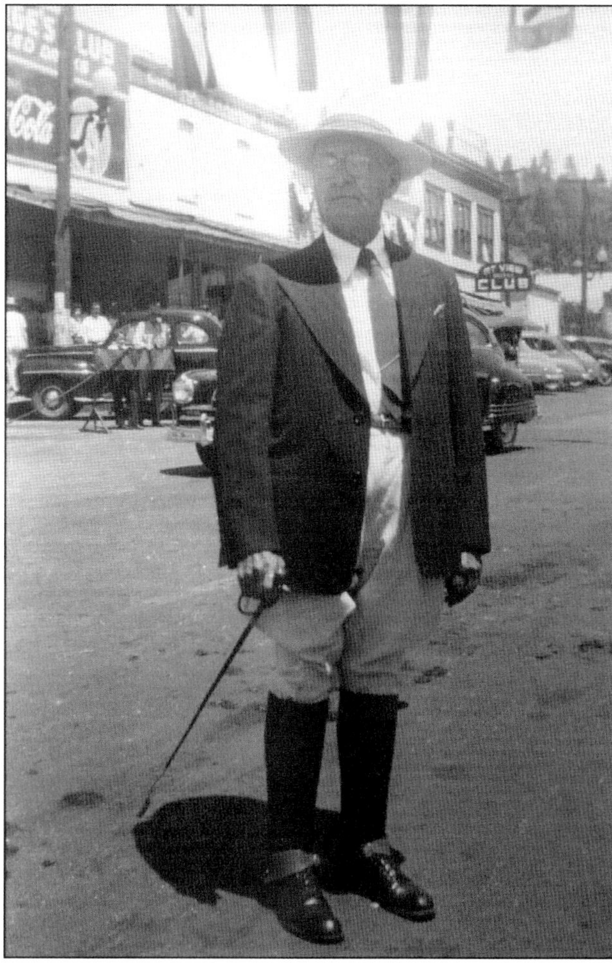

Invented in New York in the 1830s, baseball's popularity spread clear across the country in the second half of the 19th century. By the early 1900s, Placer County newspapers printed weekly game statistics for the local teams, including runs, hits, and errors. The Colfax ball club is seen below perched on the old Southern Pacific roundhouse. Some of the players include, from left to right, (first row) Art or Ed Hutchinson, Fet Sparhawk, and four unidentified players; (back row) unidentified, Art Weaner, unidentified, Lester Starr, and unidentified. (PCM.)

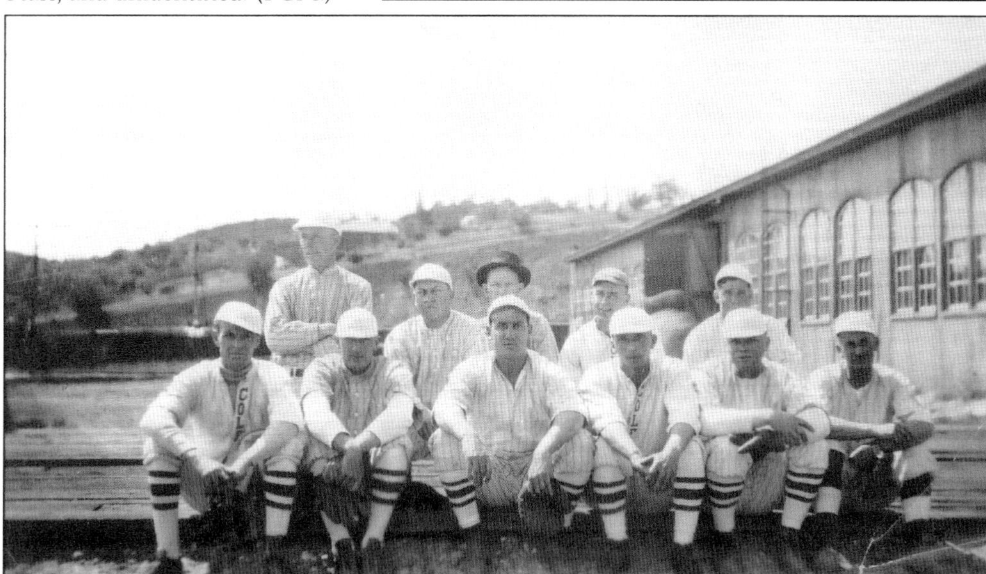

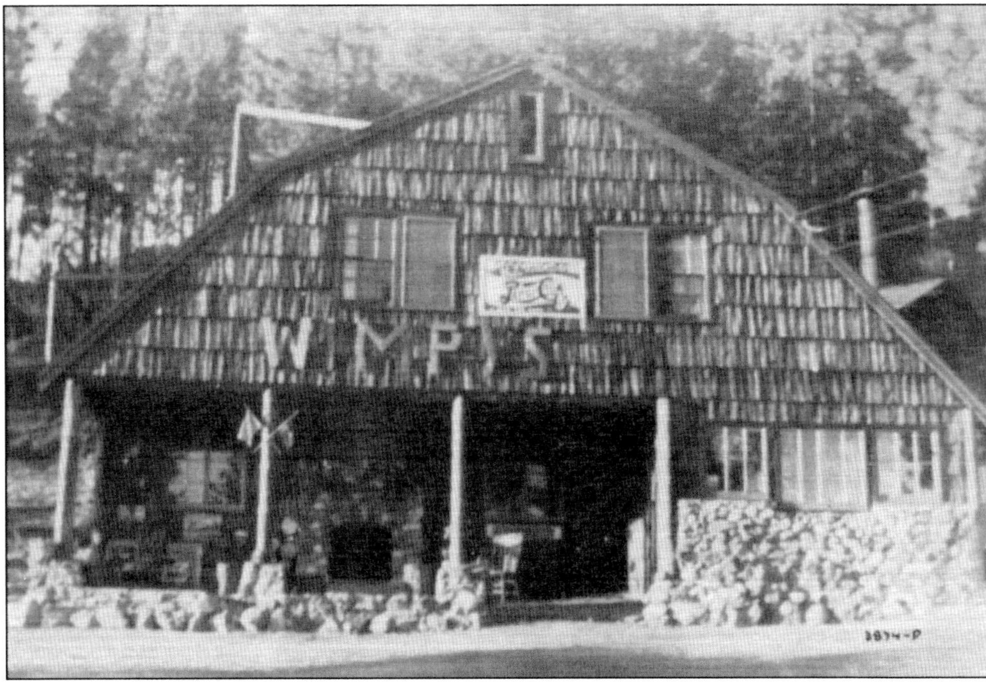

Wimpy's Fireplace was a local eating icon. Many locals remember the restaurant's namesake, Wimpy, the cartoon character from *Popeye* who was always seen with at least one or two giant hamburgers he was in the process of consuming. The restaurant became famous for its burgers. The former restaurant, now a residence, sits on the edge of Colfax on Highway 174. (AS.)

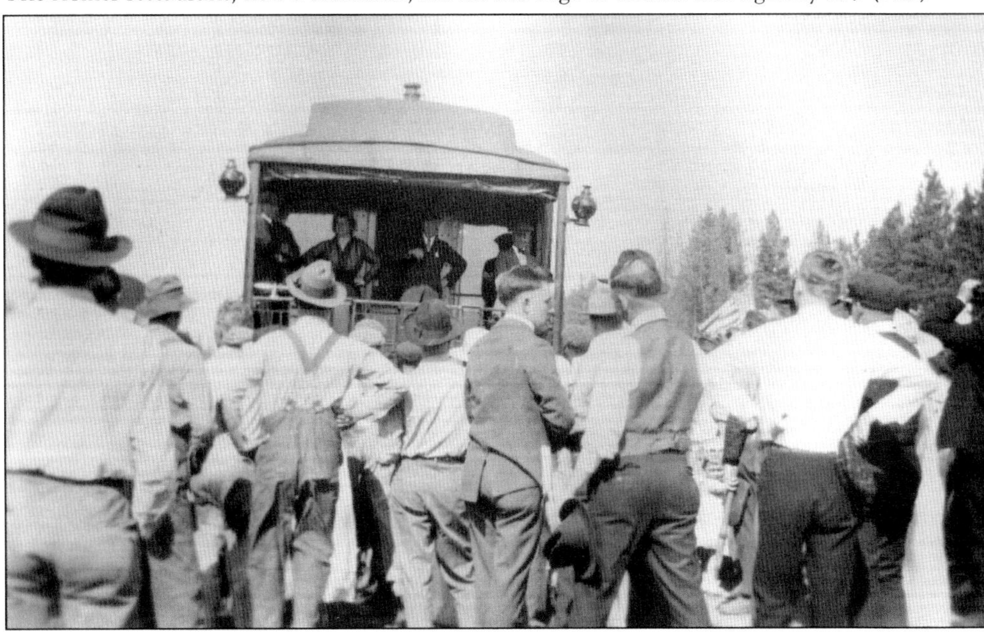

In 1918, the 24th governor of California, William Dennison Stephens (1917–1923), stopped to speak to the citizens of Colfax. He had served three terms in the US House of Representatives, from 1911 to 1916, and like many famous people of the time, he spoke from the back platform of the train. (CHM.)

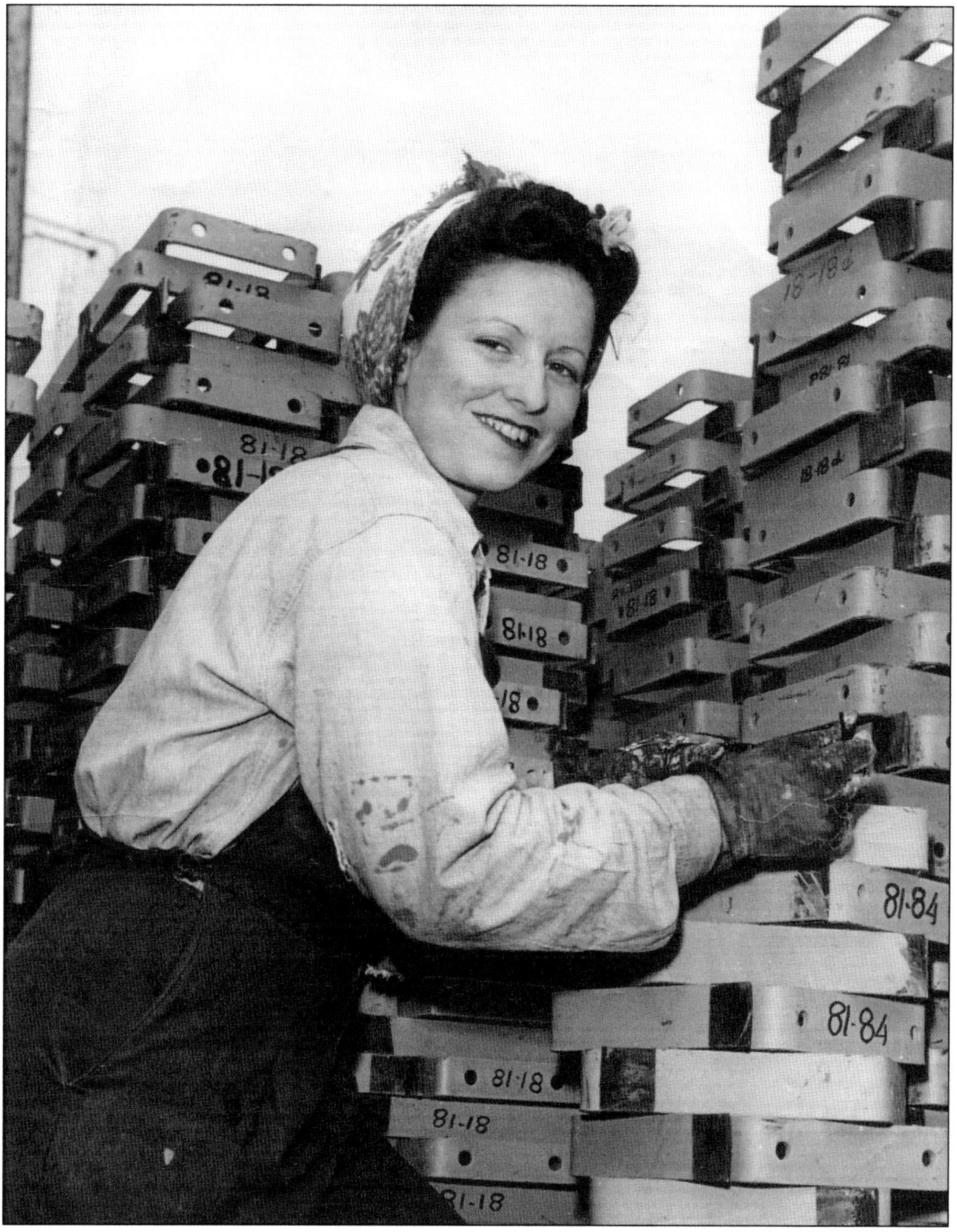

All over the nation, women stepped up to build ships and other necessary equipment. Some backfilled men in nontraditional jobs when the men went overseas, and others enlisted themselves, becoming part of the armed forces, many as translators or nurses. They learned additional skills that they may not have known before or put to full use. Naomi Prior, a longtime Colfax resident, worked at the pharmacy from 1934 to 1940. One summer during World War II, she recalls painting brackets and the interior of a Liberty ship at Marin Shipyard, near San Francisco. (CHM.)

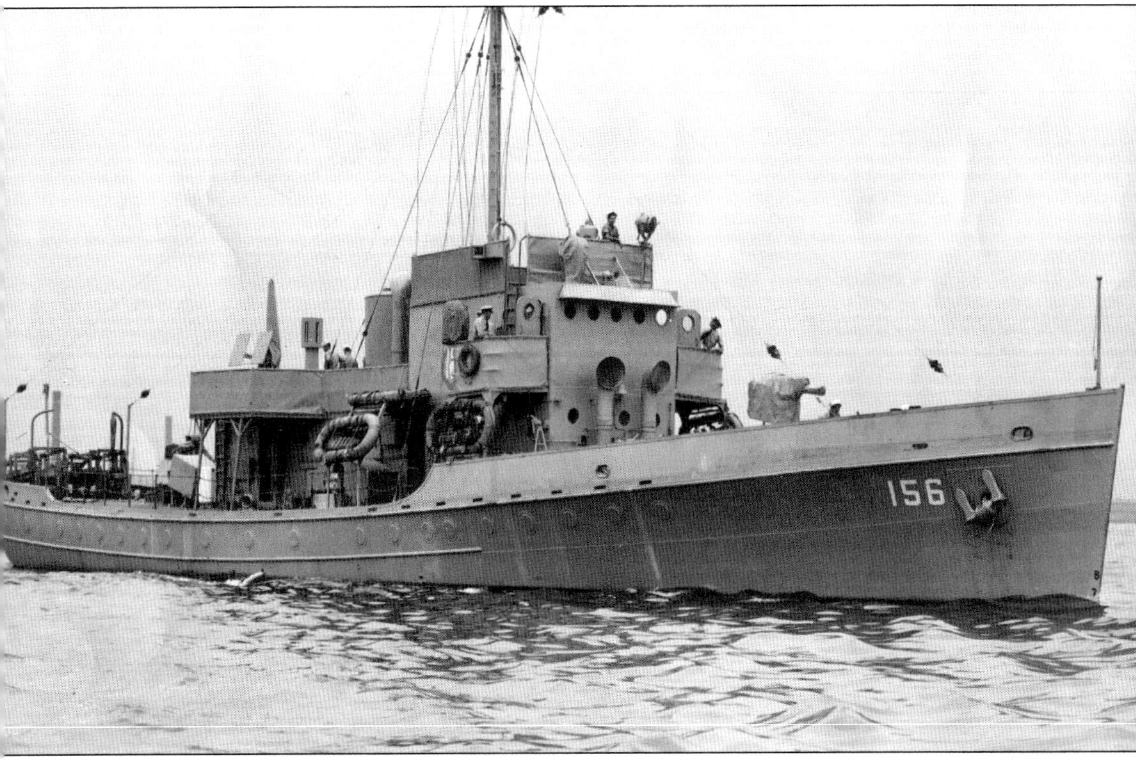

This photograph of a US Coast Guard cutter was presented to Colfax mayor Robert A. Peers on December 2, 1943. Some say that the ship, built in 1927, was named after former vice president and town namesake Schuyler Colfax. One of Colfax's cousins reportedly commented that many women were in the Coast Guard as "keepers of the light," and that the cutter was likely named after Harriet Colfax. (CHM.)

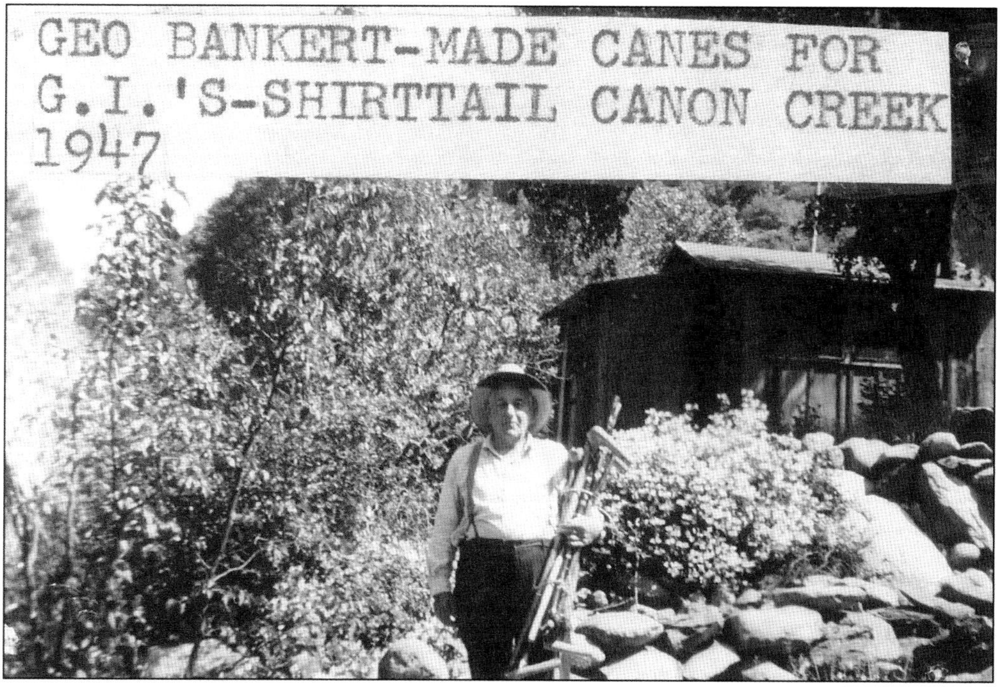

George Bankert, seen here in 1947, gathered wood, which he then carved into walking canes for World War II veterans. Bankert lived in a cabin near Shirttail Creek. A suspension bridge for foot traffic only crossed the Shirttail Canyon near the North Fork of the American River. (PCM.)

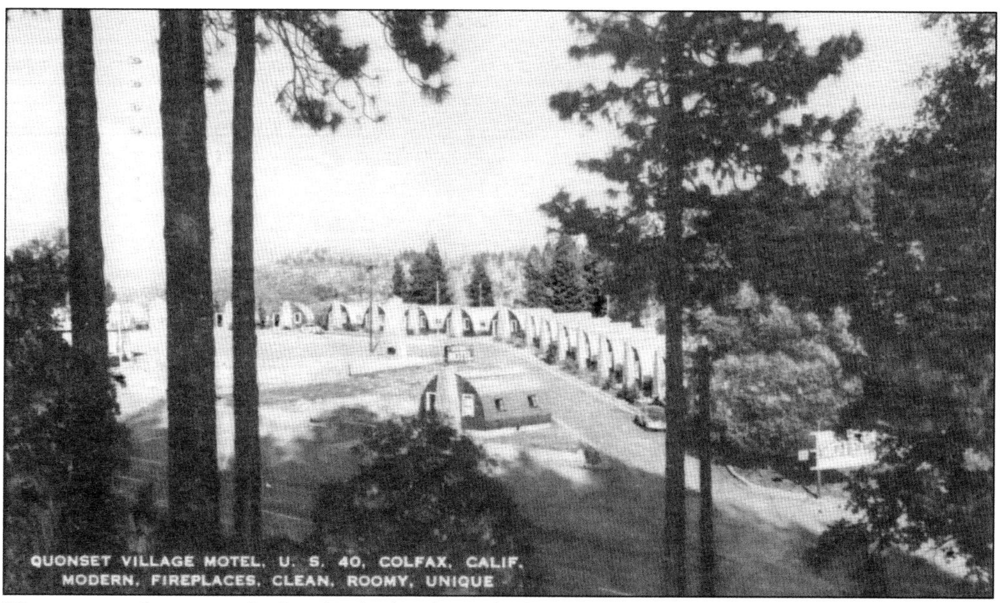

This postcard, postmarked 1953, looks through the woods at this collection of Quonset huts that served as a motel. Quonset Hut Village Motel advertised on that back that, in addition to the "clean and roomy" lodging, it offered a coffee shop and its own gas station. It was "out of the snow, yet close to winter activities." Today, only the stone skeleton of the lodge remains on Highway 174. (AS.)

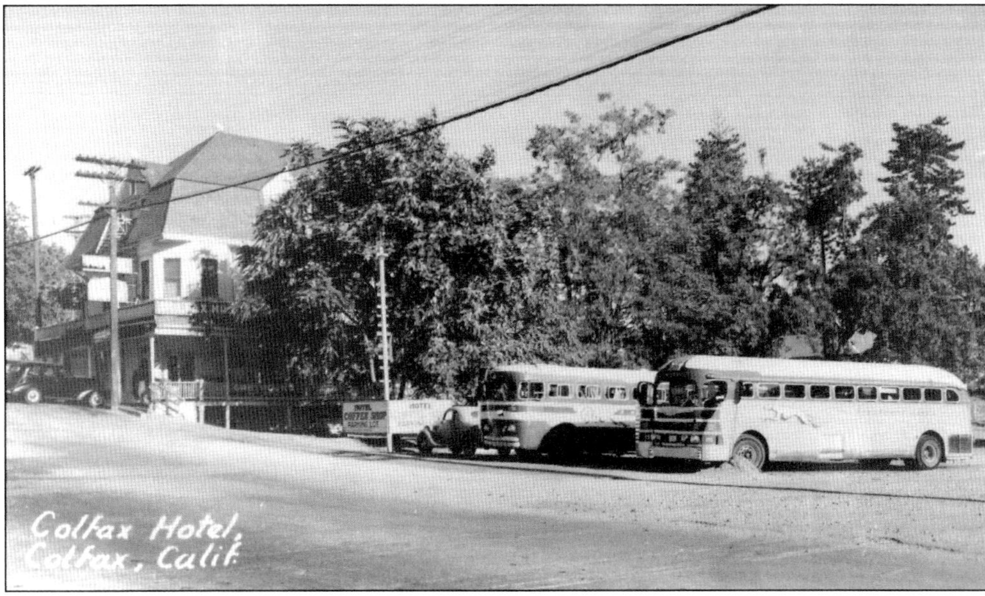

Colfax is still a transportation center for the Sierra foothills, and Greyhound bus schedules coincide with the train schedules. This photograph shows a line of waiting buses in front of the Colfax Hotel. (JLJ.)

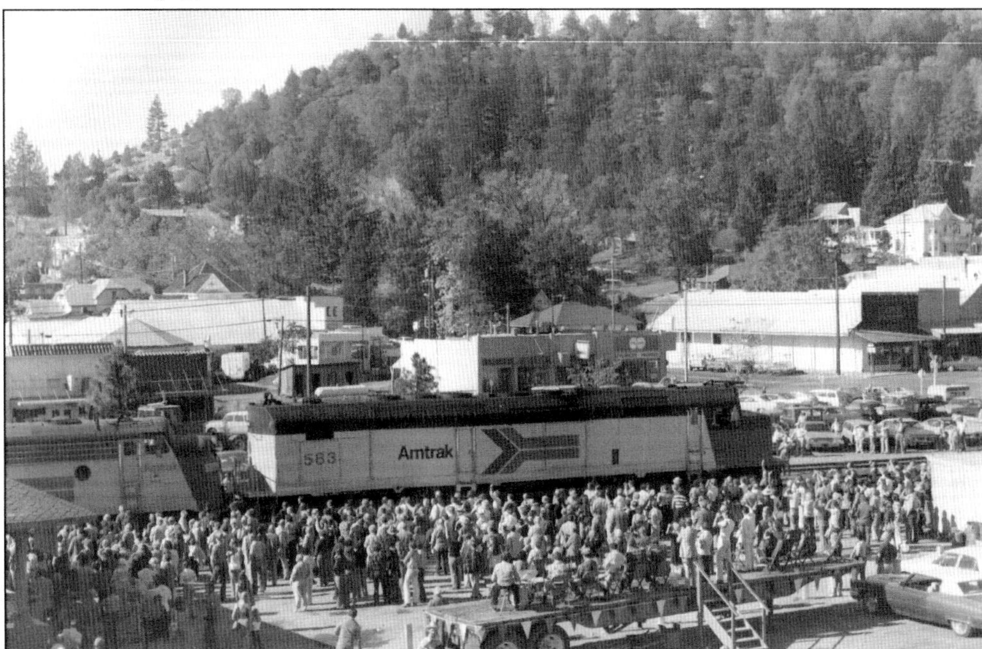

Trains, now operated by Amtrak, still come through Colfax a few times a day. Amtrak resulted from the 1971 Rail Passenger Act. As part of the National Railroad Corporation, Amtrak took over the Southern Pacific trains. Its route initially skipped the stop in Colfax, but the town protested and the train was eventually proven essential to the welfare of the town and region. A favorite trip, even for locals, is to hop aboard the train after a snowfall and travel in and out of the mountain tunnels to see the breathtaking views. (CHM.)

Six

Neighboring Communities

As gold triggered the need for mining supplies and better distribution, towns developed in the foothills to accommodate the demand. Lumber was required for commercial buildings and cabins, railroad ties, tunnels, flumes, and then for shafts after the advent of hard-rock mining. Wagons, pack trains, and eventually the railroad provided the transportation for the area's goods and supplies. Local products like timber, gold, and fruits reached worldwide markets when the transcontinental railroad connected east and west. Mail, once carried by the Pony Express, was later transported by trains and eventually airplanes.

Many small communities originally connected only by dirt roads were now linked by bridges, roads, and rails. Each contributed its strength to the local economic equation. In the central foothills surrounding Colfax, mining camps and logging towns sprouted up, either growing or gradually subsiding. Some even became ghost towns. Many miners fled the Sierra foothills in 1858, tempted by the silver rush in Nevada. Survival was largely based on each town's assets: gold, timber, tillable land, and train depots.

Many of the miners who stayed on eventually turned to other trades. Communities established the necessary infrastructure beyond saloons, hotels, and assay offices. Colfax's neighbors include Alta, Applegate, Chicago Park, Clipper Gap, Dutch Flat, Gold Run, Iowa Hill, and Weimar. Each offers its own personality and history.

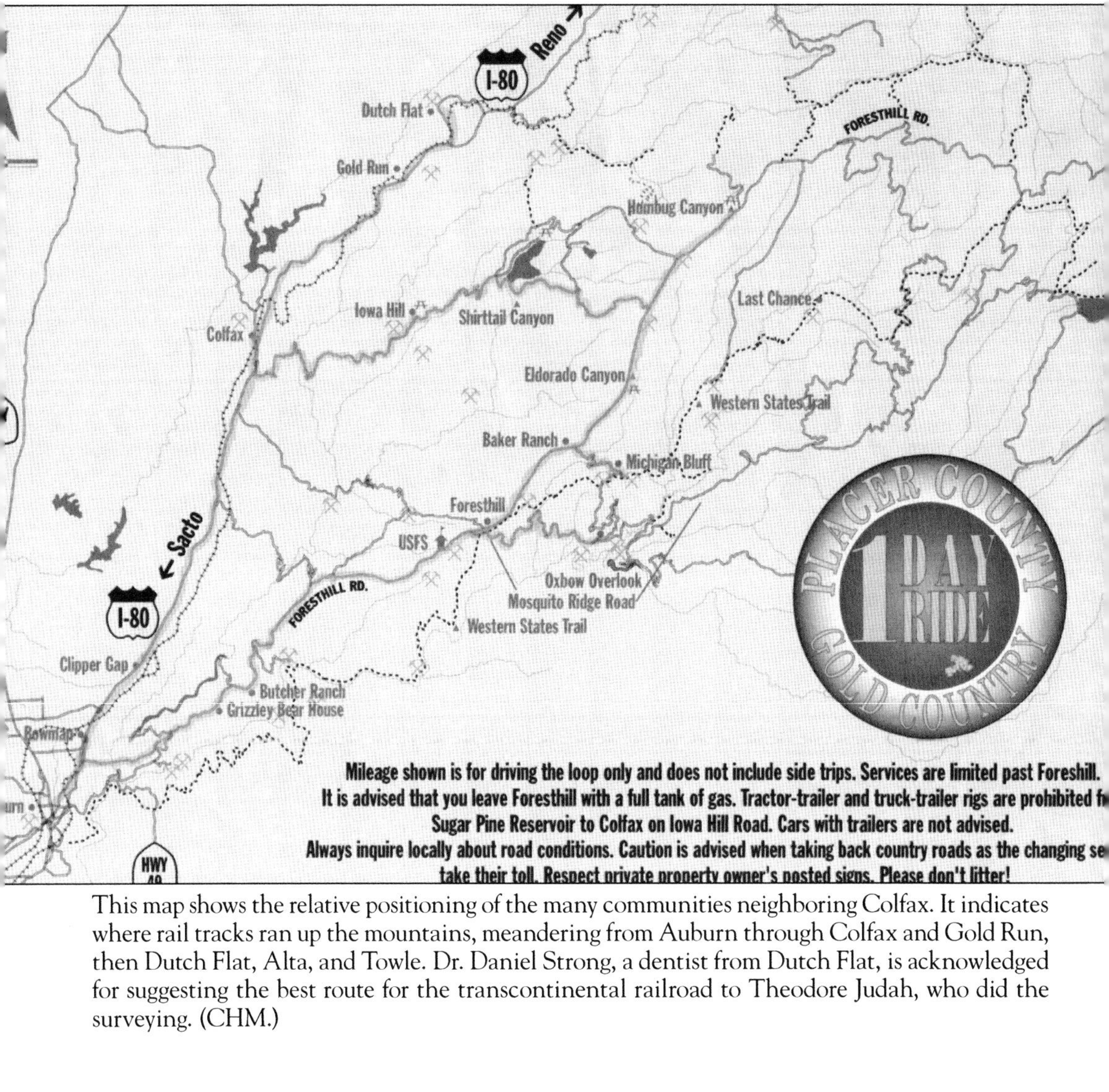

This map shows the relative positioning of the many communities neighboring Colfax. It indicates where rail tracks ran up the mountains, meandering from Auburn through Colfax and Gold Run, then Dutch Flat, Alta, and Towle. Dr. Daniel Strong, a dentist from Dutch Flat, is acknowledged for suggesting the best route for the transcontinental railroad to Theodore Judah, who did the surveying. (CHM.)

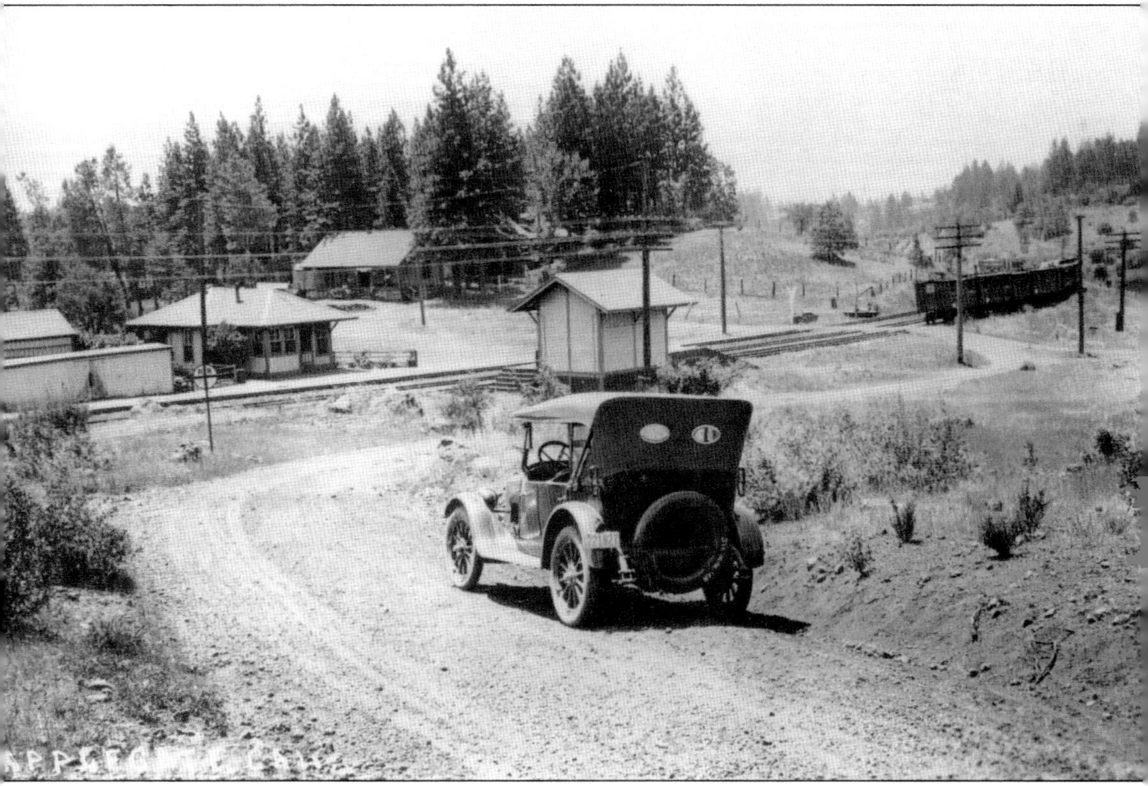

Applegate sidled the road between Auburn and Illinoistown on the Bear River. The town was established in 1849 and was originally called Lisbon after early resident Lisbon Applegate. At one time, it even went by the name Bear River House. In 1855, G.W. Applegate became the first postmaster of Applegate. The Central Pacific Railroad built two stations, as the tracks were separated by about a quarter-mile. The station serving eastbound travelers became East Applegate. Applegate still exists as an unincorporated town. (AS.)

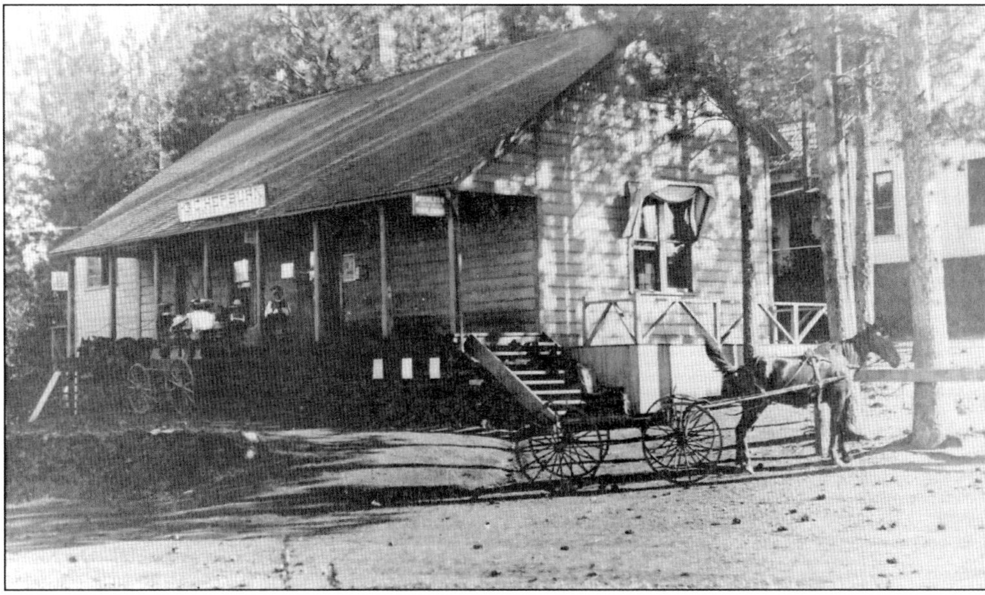

This photograph of the C.C. Hepburn store in Applegate is dated by the horse-drawn cart in the center and the buggy full of ladies and one young lad in front of the store. (JLJ.)

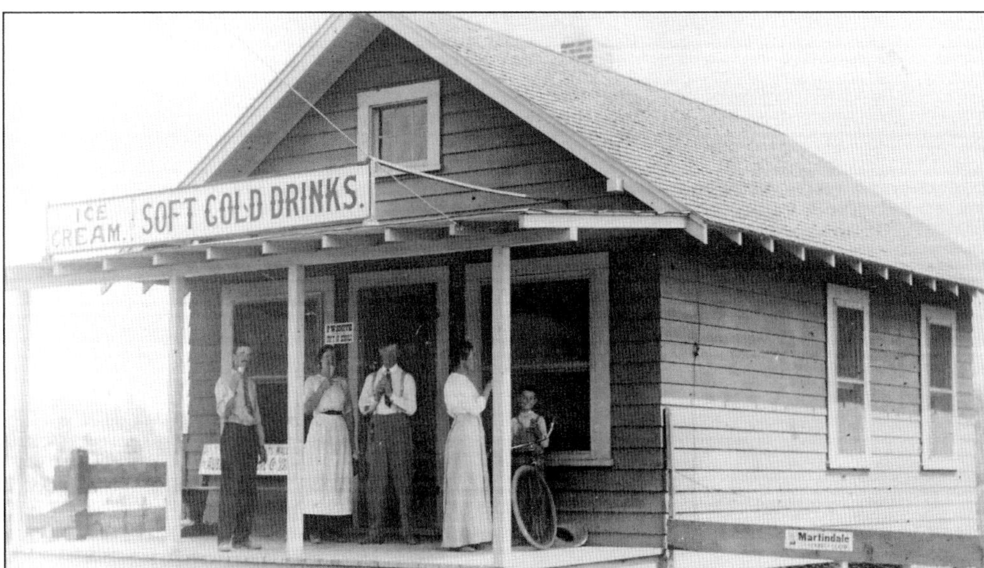

This little store in Applegate offered ice cream cones to cool off on a hot day. This small crowd gathered on the front porch to enjoy the refreshment in 1914. (JLJ.)

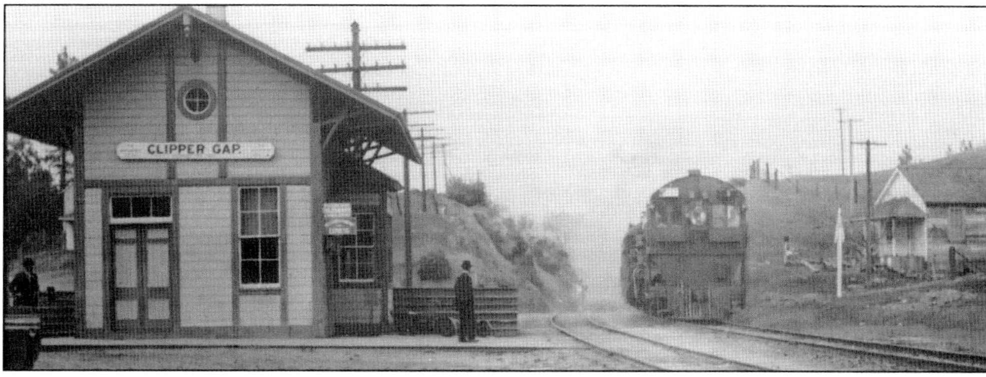

The Clipper Gap post office was established in 1866 for this small community, which did not become known as Clipper Gap until 1894. The trains stopped at the local depot as part of a regular schedule from Sacramento beginning in June 1865. Clipper Gap, an unincorporated community, lies 6.6 miles northeast of Auburn and about 11 miles southwest of Colfax. (AS.)

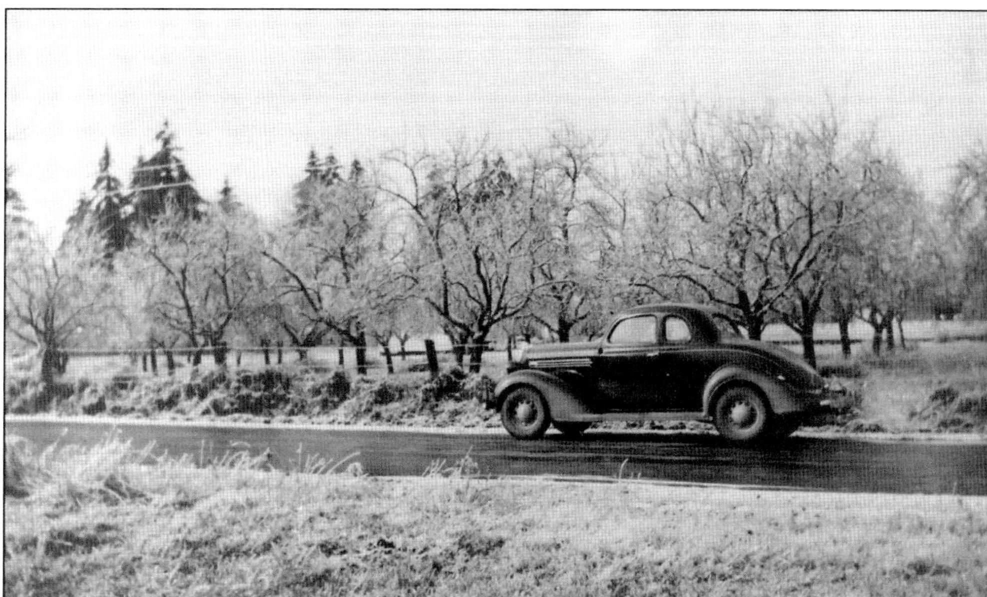

This scene in Chicago Park shows the prolific fruit orchards that still line Highway 174 between Grass Valley and Colfax. During the days of the Nevada County Narrow Gauge Railroad, there was a train depot in Chicago Park where passengers got on and off and fruit crates were loaded for transport to Colfax. The local train linked up with the transcontinental railroad for travel east and west across the country. Since 1902, the Bierwagen family has grown fresh fruit in Chicago Park. The homestead is still in operation as the Happy Apple Kitchen and store, offering fruit, lunch, and pies to those traveling along Highway 174. (JLJ.)

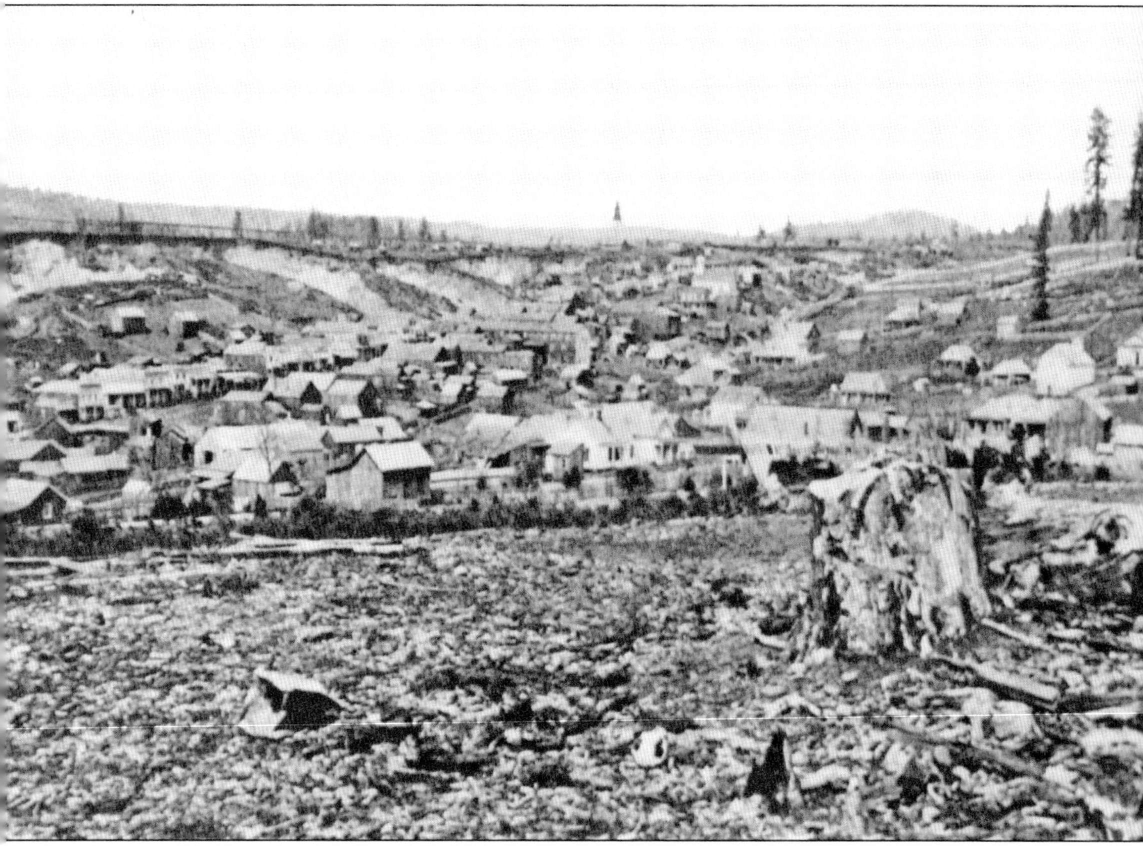

In 1851, Joseph and Charles Dornback came from Prussia (now Germany) and founded a new town 13 miles north of Colfax. Muleskinners called it Dutch Charlie's Flat and later just Dutch Flat. This early photograph shows the ruddy dirt streets, wooden buildings, and in the background, the hydraulic mining. The Central Pacific Railroad construction of the transcontinental railroad progressed just a short distance from town. Before the railroad stretched from Sacramento over the Sierra Nevada Mountains, pack trains delivered supplies for Dutch Flat. Stagecoaches also made regular trips with passengers and their meager belongings. By 1860, Dutch Flat supported a church, a newspaper, a school, three hotels, and numerous saloons and other businesses. (DFL.)

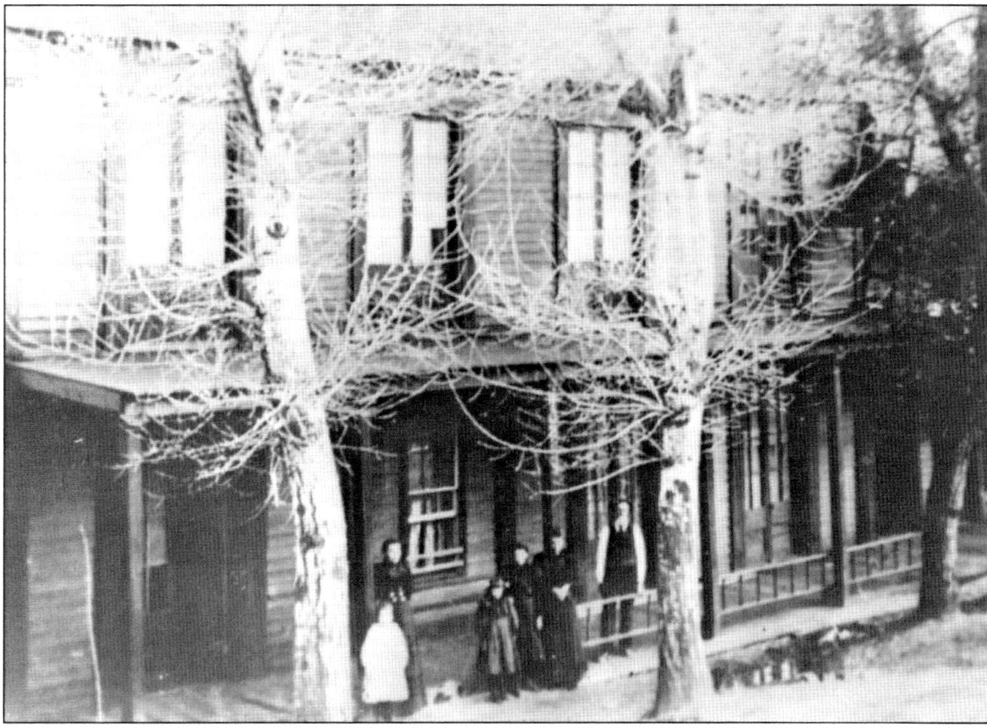

The opera house in Dutch Flat hosted amateur shows, musical performances, guest speakers—including Mark Twain—and grand balls. On July 5, 1880, when the town celebrated Independence Day, tickets were sold for $2.50, which included supper. The Aaron A. Ferguson family owned the opera house from 1872 until fire destroyed the building in 1913. (JLJ.)

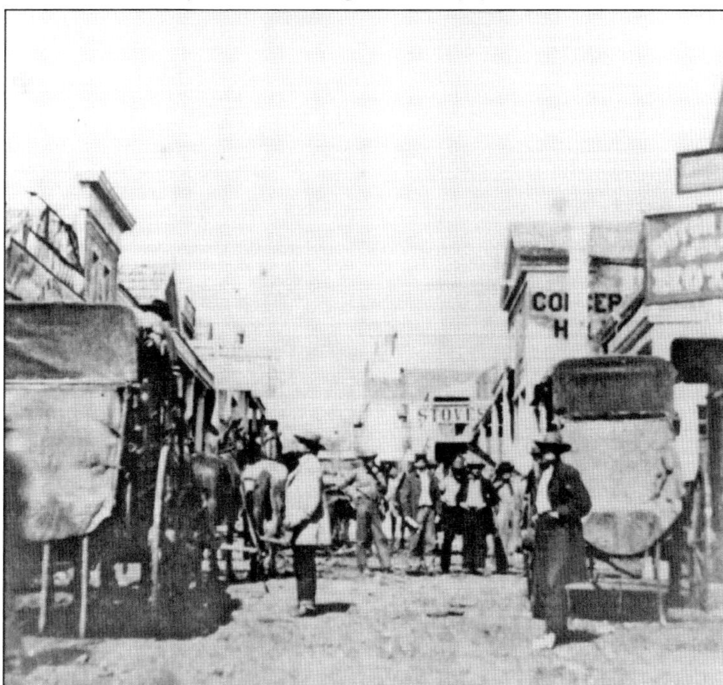

One of the richest gold-mining towns, Dutch Flat once boomed with activity as stagecoaches passed through carrying passengers up and down the mountain. Businesses sprang up in boom times to meet the growing demand for goods, lodging, and entertainment. The Dutch Flat Hotel, on the right, still stands today at the corner of Main and Stockton Streets. Note all the men standing tough in the middle of the street sporting hats and long jackets. (PCM.)

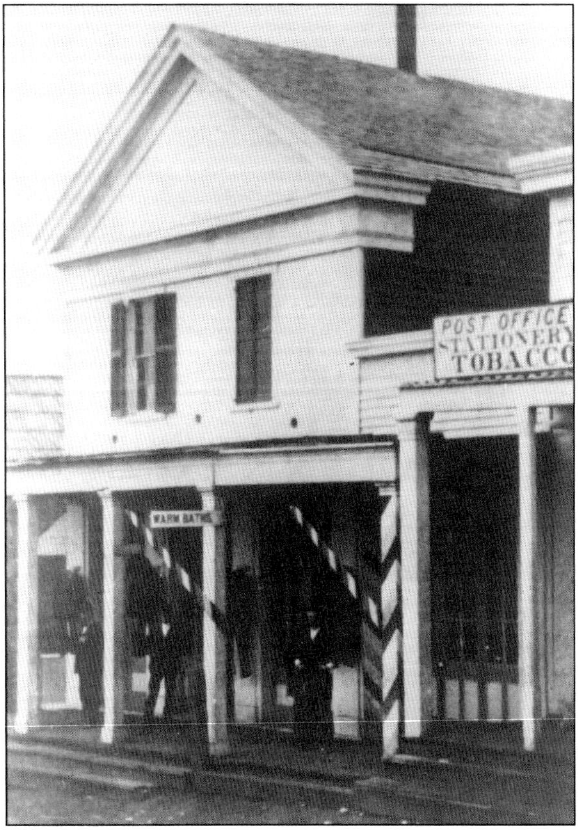

The Masonic hall in Dutch Flat is seen here in the 1860s. The barbershop on the ground floor (central building) offered warm baths, for a fee, to the public. Next door, a multipurpose business served as a post office, a stationery shop, and a tobacco vender. (PCM.)

The Odd Fellows and Masonic halls (below) began as one-story buildings but received second stories in the late 1860s (Masons) and early 1870s (Odd Fellows). The Odd Fellows met on the second floor, and a general mercantile operated downstairs. The Masons' building also housed a business on the first floor. Both buildings are still used by the organizations today. Town lore says that when the second-story floor sagged three inches, members merely shortened the back legs of their chairs to put everyone level. (JLJ.)

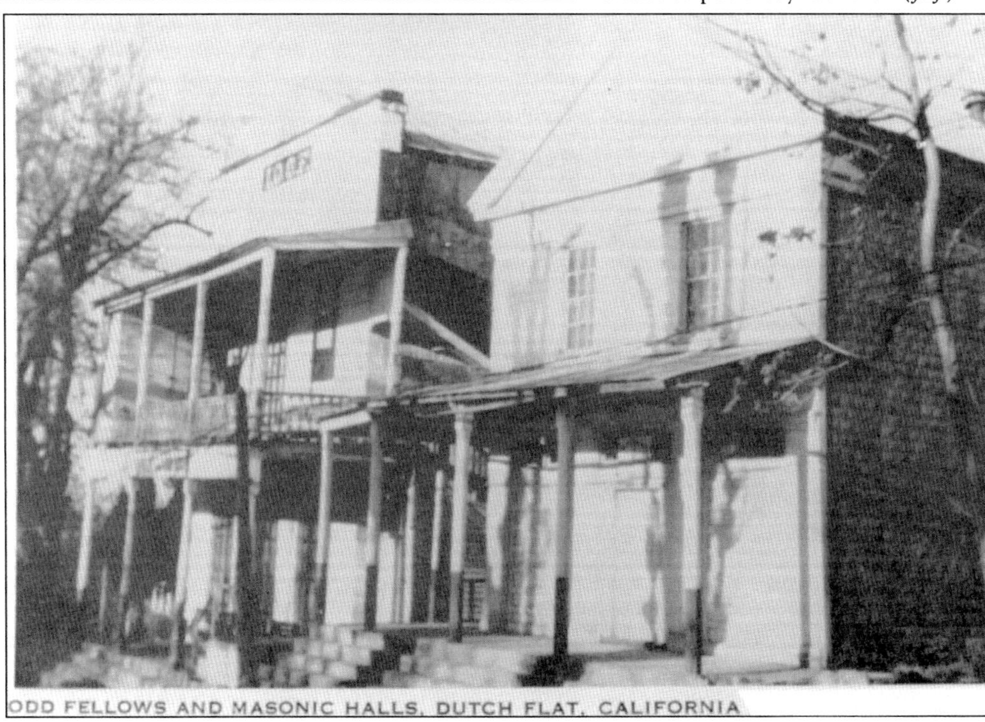

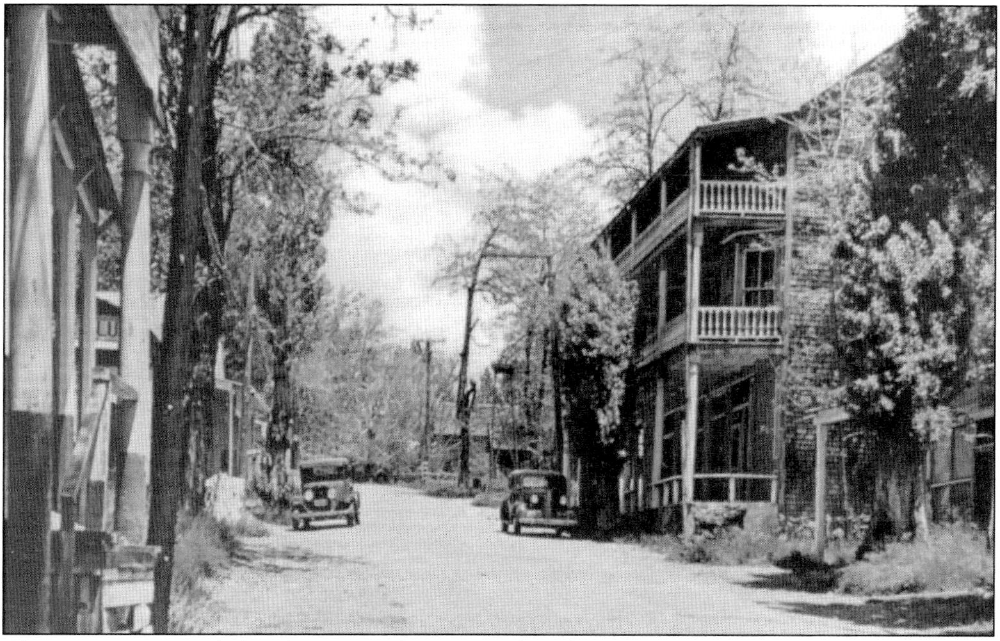

The town supported the Placer Hotel, Union Church, Dutch Flat Hotel (still standing), the Golden Eagle Saloon, and the two-story Towle Brothers' building along with other businesses long gone. The Towles owned the largest local enterprise, the lumber mills. The town continued to thrive after the Gold Rush, supplying timber for the construction of the railroad. The classic automobiles indicate that this photograph dates from the 1930s. (JLJ.)

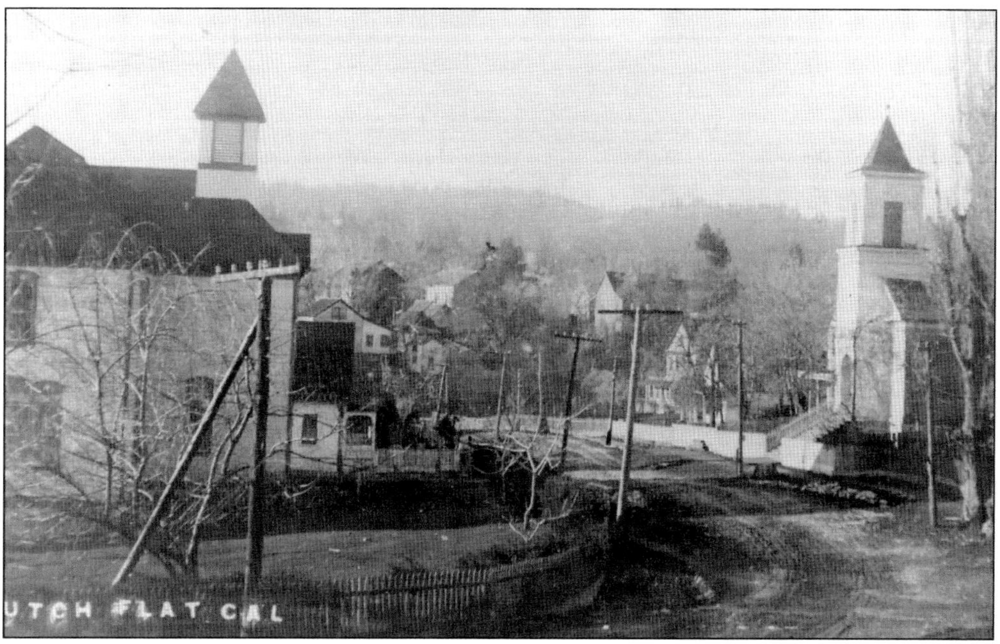

This historical scene shows the Dutch Flat School, built in 1898, on the left side of Stockton Street tucked inside the white picket fence. The Methodist church stands on the right. Tall steeples top both structures, with the church distinguished by its cross high up on the peak. Both buildings remain today, and the school now serves as a community center for the town. (JLJ.)

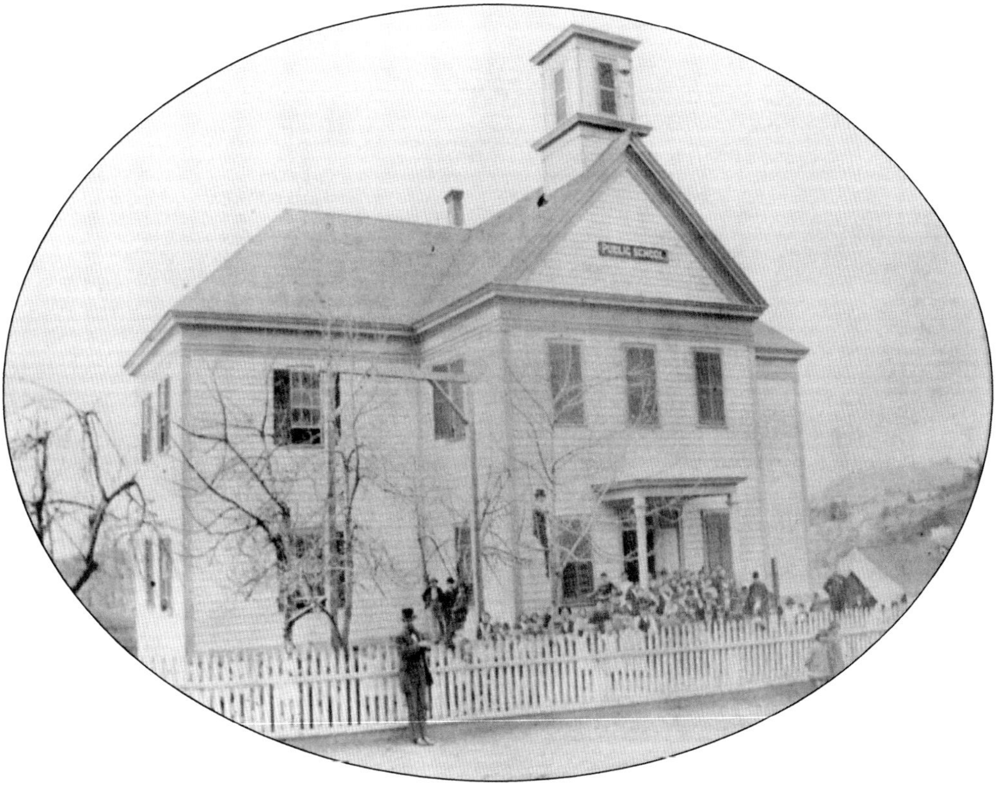

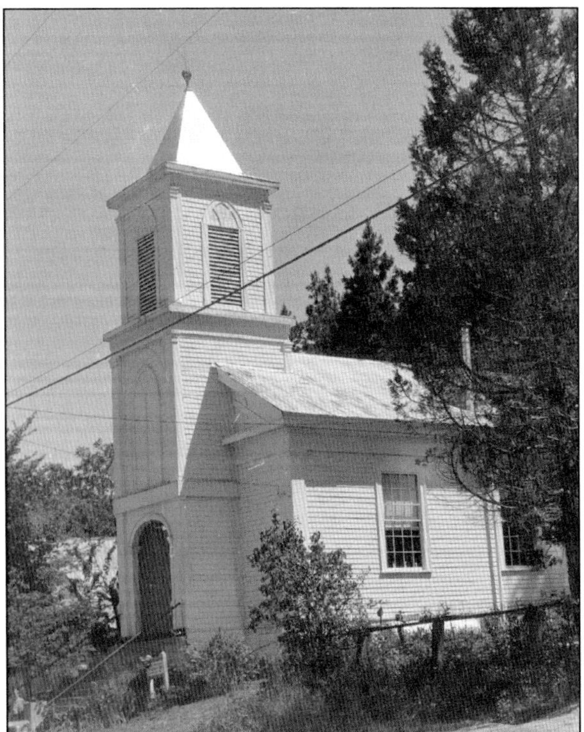

The public school building above was one of four schools for local children. The first, a one-room building on Main Street, offered classes from 1856 to 1859. From 1859 to 1875, another one-room school on Stockton Street took its place. Two two-story buildings then took their turns. The old grammar school, which operated from 1898 to 1962. In 1964, the building became the community center. Children now attend Alta Dutch Flat School. (PCM.)

The Methodist Episcopal church on Stockton Street was formed in 1857. Local citizens constructed the building in 1861, not long before the Civil War began. Some members with strong Union ties left the Methodist church and started Union Church. Today, services are still offered in the original Methodist church. (PCM.)

After some members split off from the Methodist church to form Union Church, they erected their new church building on Main Street. There was also a little wooden Catholic church between Park Avenue and Clark Street. Neither Union Church nor the Catholic church remains today. The latter burned in the summer of 1898. A schoolhouse burned in the same week. (PCM.)

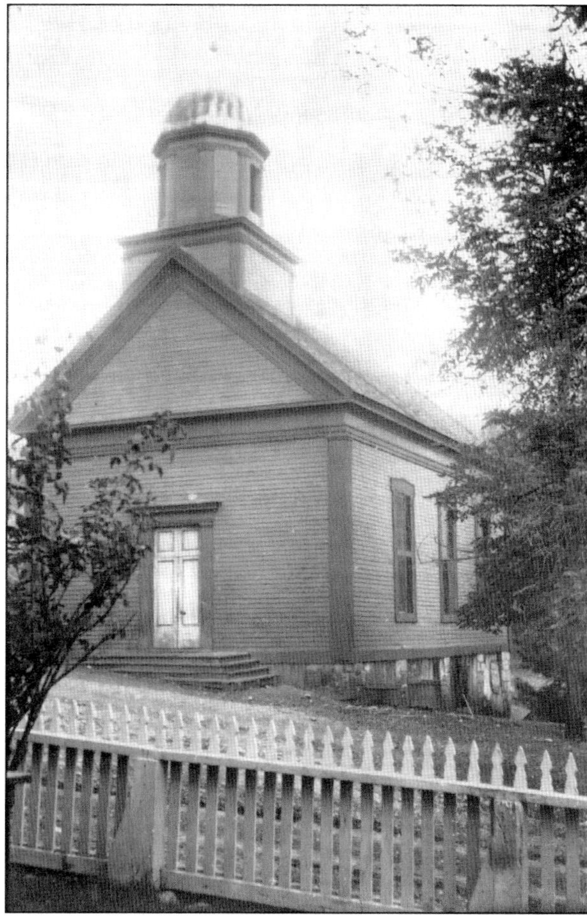

Below, folks watch on the wooden platform of the Dutch Flat railroad station as a passenger train pulls away on the tracks on Independence Day in 1866. From here, the Central Pacific Railroad worked its way up into the mountains. Construction kept on going up and over the summit through winter storms and fierce conditions. (AS.)

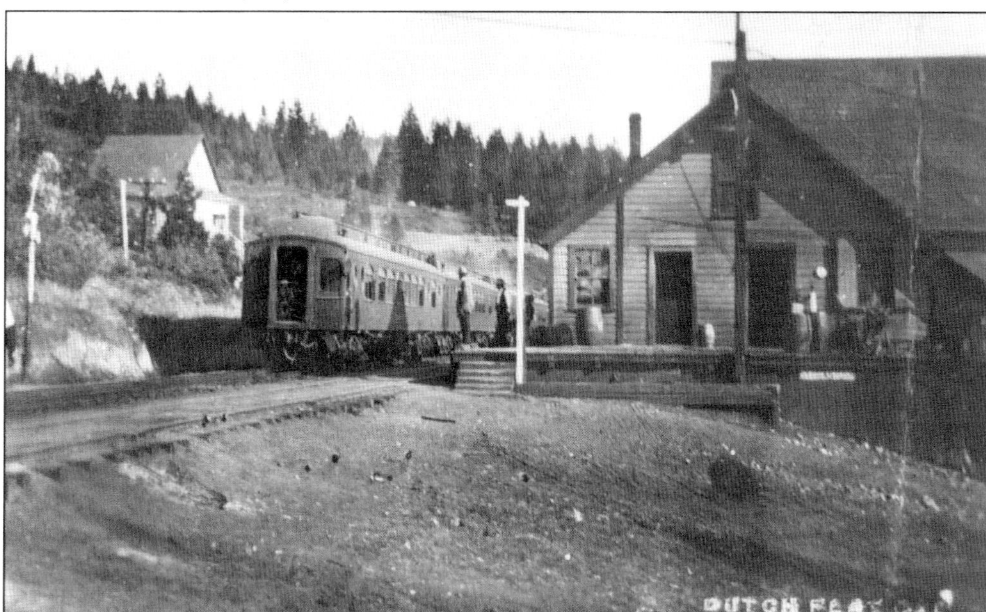

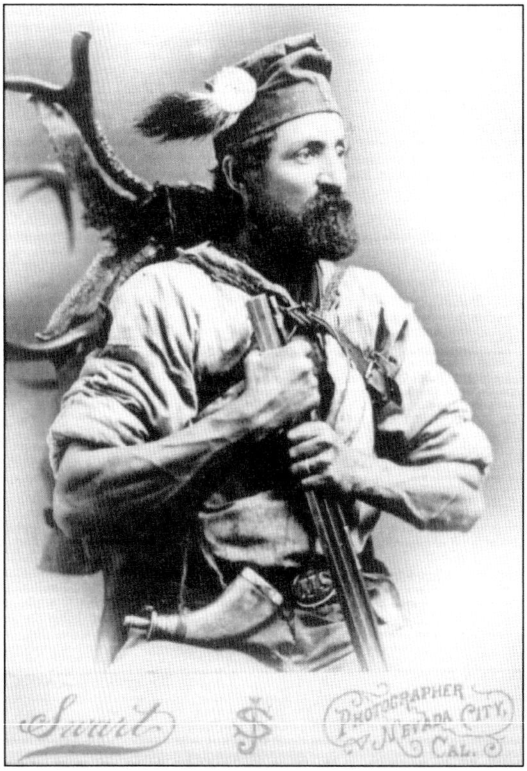

Ward Fay served as a scout, a critical role for the nation in the 1800s. Scouts gathered information about enemies, blazed trails in the west, and guided army troops as they pressed through the mountains. Fay died in Dutch Flat in 1911 and is buried in the Masonic Cemetery there. Even though he was not a family member, he lies in the Harriman plot because he was a very close friend of Samuel Harriman. (PCM.)

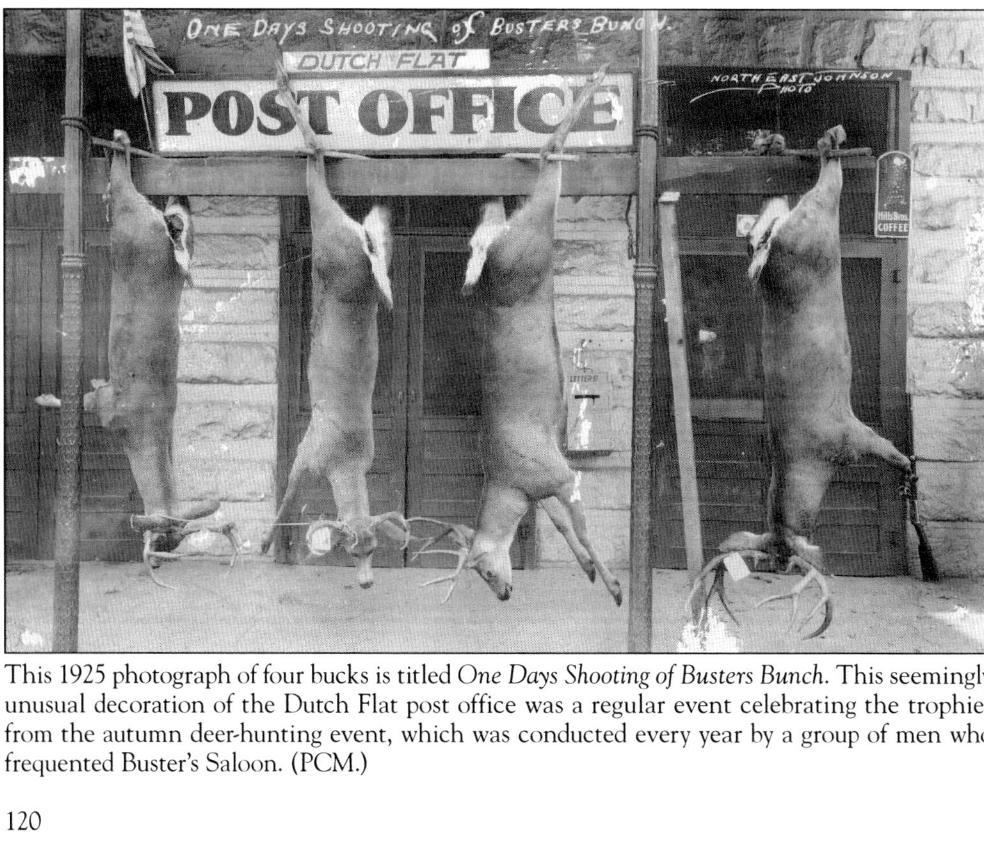

This 1925 photograph of four bucks is titled *One Days Shooting of Busters Bunch.* This seemingly unusual decoration of the Dutch Flat post office was a regular event celebrating the trophies from the autumn deer-hunting event, which was conducted every year by a group of men who frequented Buster's Saloon. (PCM.)

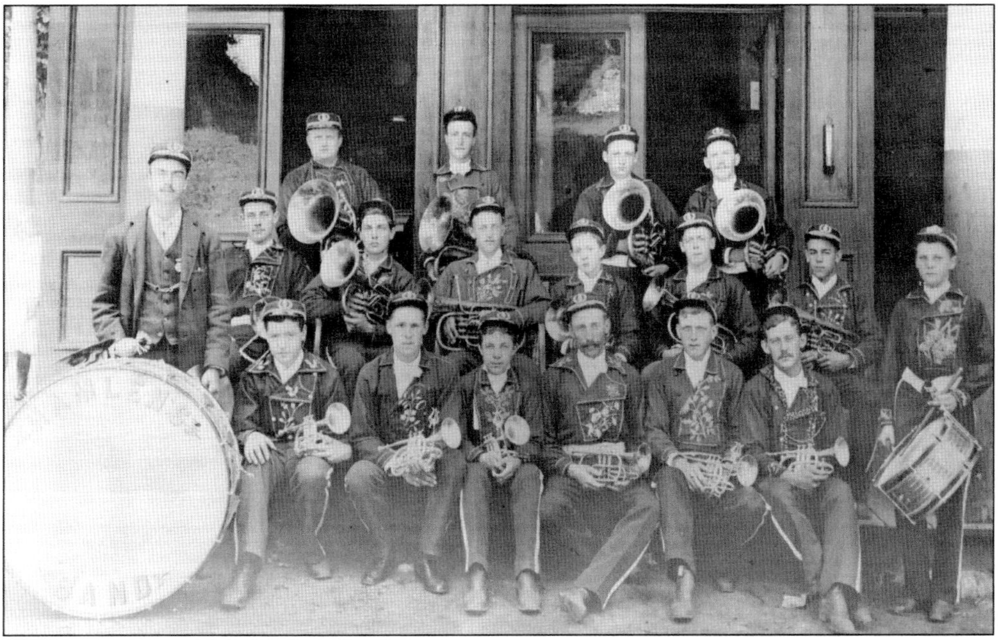

Hamlen's Band is seen above in front of the hotel in 1895. Members included, from left to right, (first row) Arthur W. Nicholls, Frank La Kamp, "Wally" Nicholls, Cal Hamlen, George Knapp, Jay L. Waggoner, and Ed Bellows; (second row) Fred Hudson, Phil A. Niles, Frank Shaffer, Henry Olson, Loup Faller, Glen Wedgewood, and Nelson Waggoner; (third row) Andrew Baker, George Gray, Edward Gray, and George Johnson. Cal Hamlen's brother owned the local butcher shop in town. (PCM.)

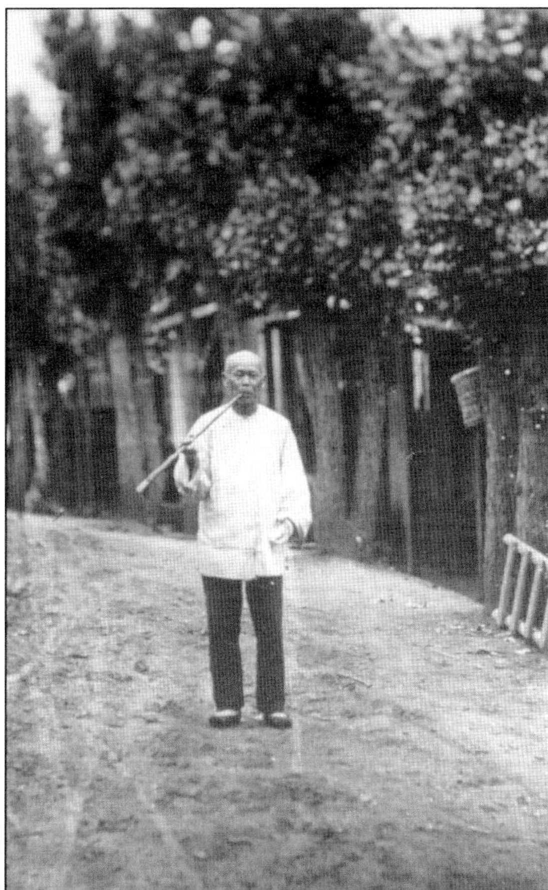

News touted 2,000 Chinese in Dutch Flat between 1851 and 1873—more likely less than half that number. After building the transcontinental railroad, some established local businesses in town, and others engaged in mining. There were actually two different Chinatowns in the community. After the first one burned, another was built on the hill. This photograph shows an elderly Chinese man standing alone smoking his long pipe flanked by the ramshackle buildings in the old Chinatown in Dutch Flat. (PCM.)

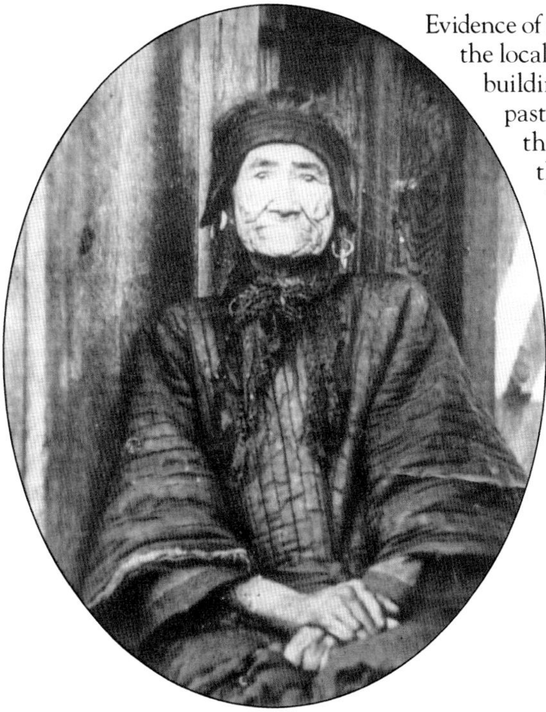

Evidence of the Chinese populace in Dutch Flat is seen in the local Chinese cemetery. An adobe rammed earth building, which still stands on Sacramento Street past Stockton Street, served as a storehouse for the Chinese population in about 1877, after their neighborhood burned to the ground. The lady seen here, known as China Mary, was well known as a professional wailer for funerals. She traveled great distances from her home in Virginia City—as far as Dutch Flat and Grass Valley—and was paid by the grieving family to give a eulogy and to sing mournful songs at the ceremony. (PCM.)

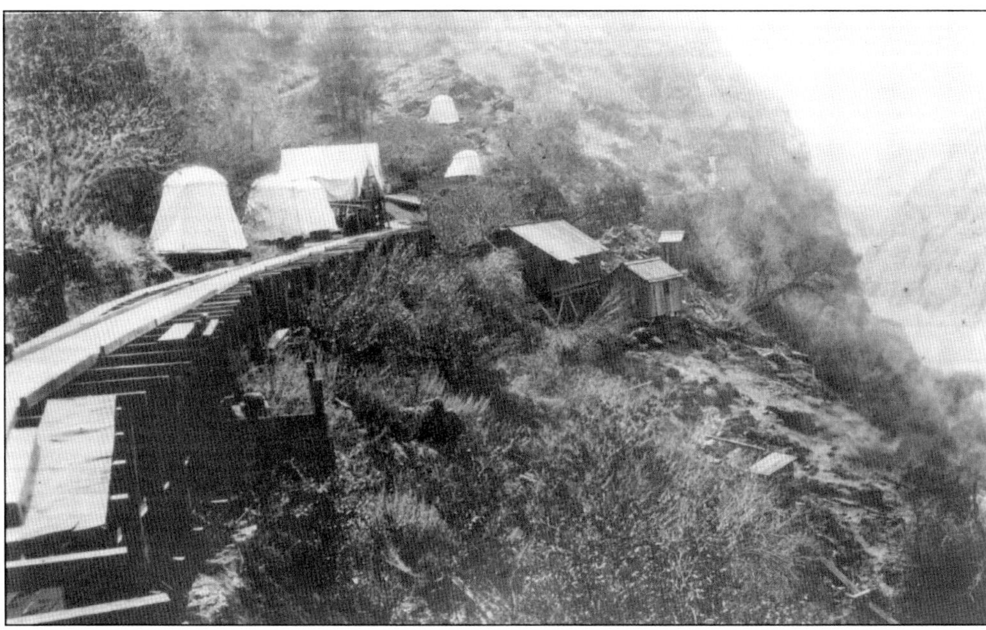

Tents, such as these in the photograph near Dutch Flat, were pitched along the route of the transcontinental railroad during its construction. Come winter, these shelters provided little comfort or warmth to the mostly Chinese workers who inhabited the tents on the tenuous slopes. Chinese employees had to supply their own tents. The infrastructure to build this rail system through the rugged Sierra Nevada Mountains required millions of lumber feet. Luckily, the Dutch Flat area had a great lumber resource, the Towle brothers' company. The brothers operated their own private narrow-gauge railroad to transport their lumber and supplies. (PCM.)

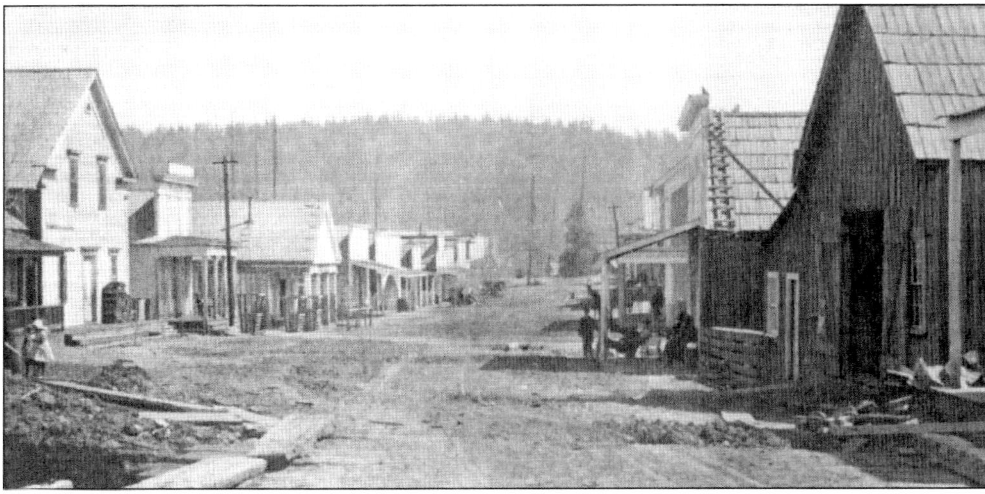

This town, originally called Mountain Springs opened its post office in 1854. The name changed to Gold Run in 1863 when the post office relocated one mile to the north. Over $6 million worth of gold was mined there using hydraulic mining between 1865 and 1878. Mining operations closed down in 1882 following the court ban on hydraulic mining. The center of town, seen here in 1868, had peaked-roof buildings with wooden shingles. Here, men leisurely sit in front of a business, one tilted back in a chair. This settlement is no longer a town, but still populated, and Landmark No. 405 off of Interstate 80 denotes its historic role. (PCM.)

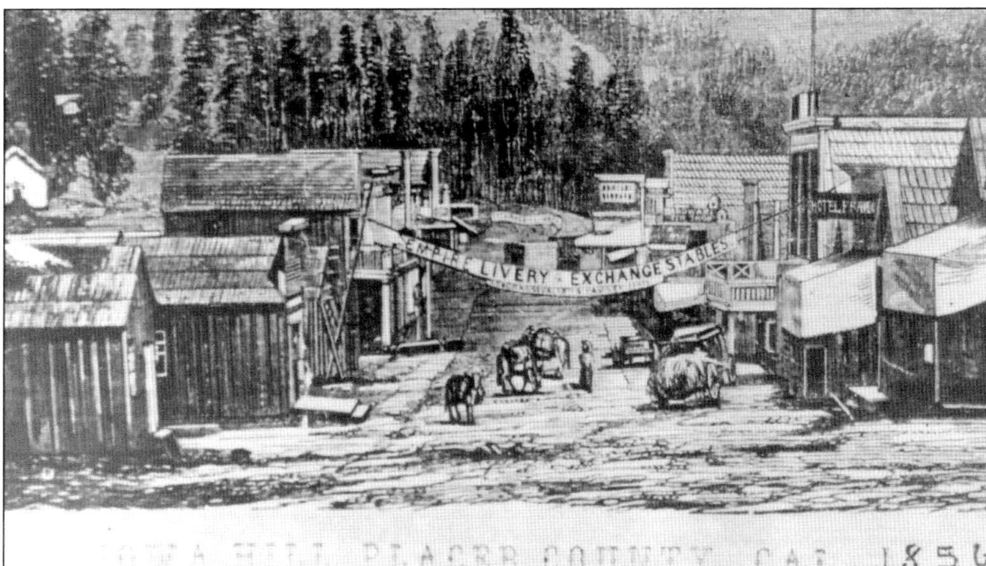

This Iowa Hill drawing typifies an 1850 California gold-mining town, complete with wooden buildings, peaked roofs, packhorses, and wooden-wheeled wagons. The banner across Main Street promotes the Empire Livery and Exchange Stables. In 1853, the first gold was discovered at the Kennedy claim, in a gulch entering the North Fork of the American River. The North Star became the other primary claim. The town, built on the ridge between the American River and Indian Canyon, burned in 1857 but was quickly replaced. It offered numerous stores, three groceries, four hotels, a school, a theater, two fraternal lodges, two churches, and the usual billiard halls and beer saloons. A devastating fire occurred again in 1922, and the town never regained its prosperity. (JLJ.)

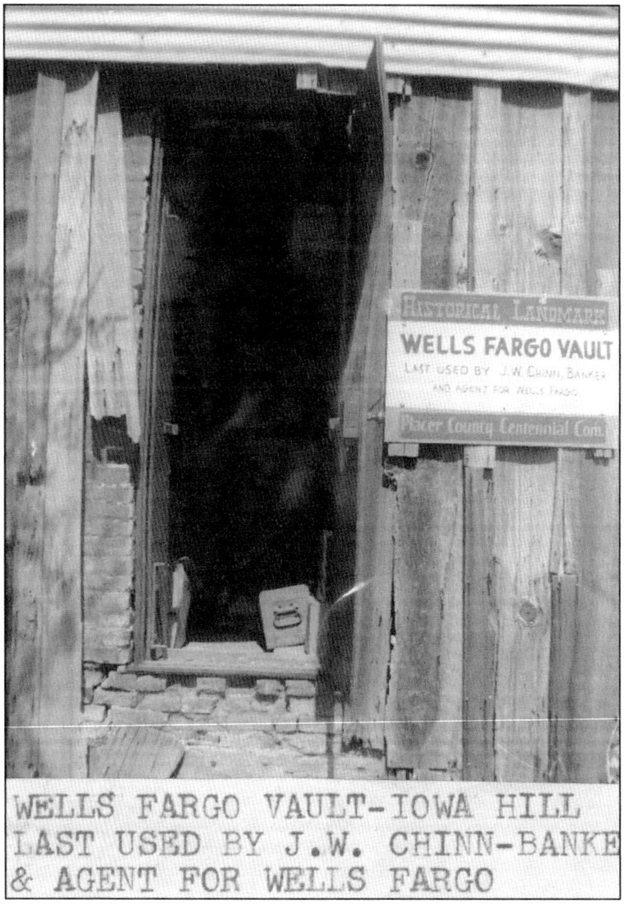

The remains of the Wells Fargo building in Iowa Hill are seen on this page. The business dates back to 1855 and is now a historic monument. The metal door offered security over the years, protecting the gold and money that came and went from the vault. (PCM.)

By 1948, when the photograph below was taken, the vault had been reduced to its brick superstructure. The last banker and agent for Wells Fargo here was J.W. Chinn. (PCM.)

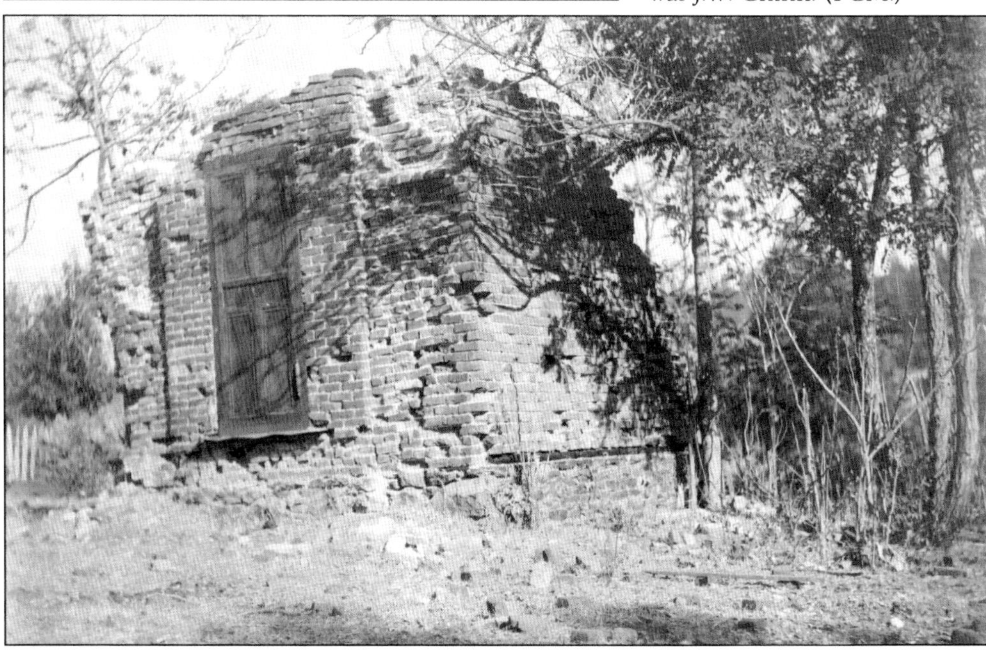

The road to Iowa Hill is known for its twists and turns and steep drop-offs. Some traveled by wire to go up to Iowa Hill for groceries. Sniper's Crossing stretched over the North Fork of the American River about a mile below Pickering Bar. On the cable, Ed Martinson pulls himself along, seated on a swinging bar. (PCM.)

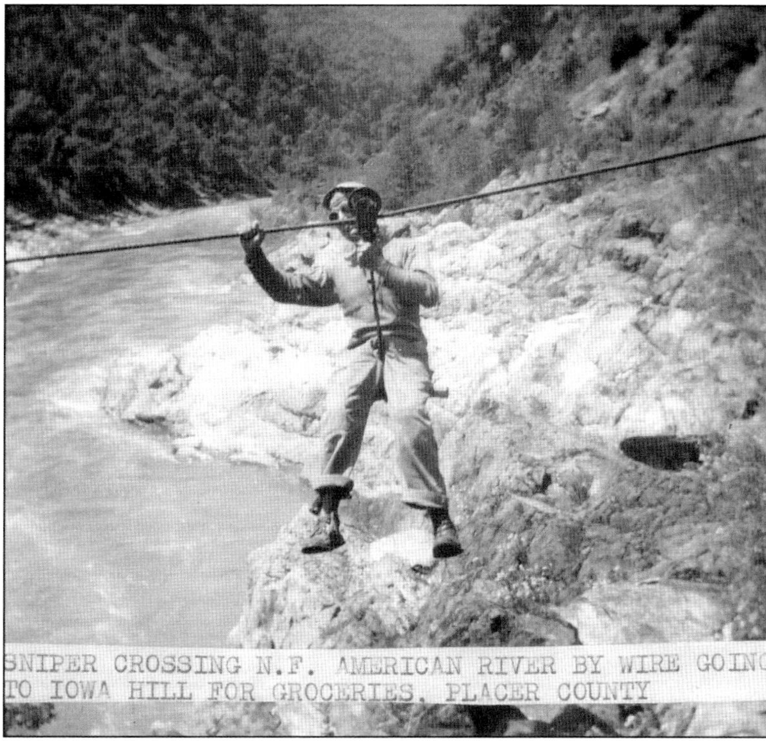

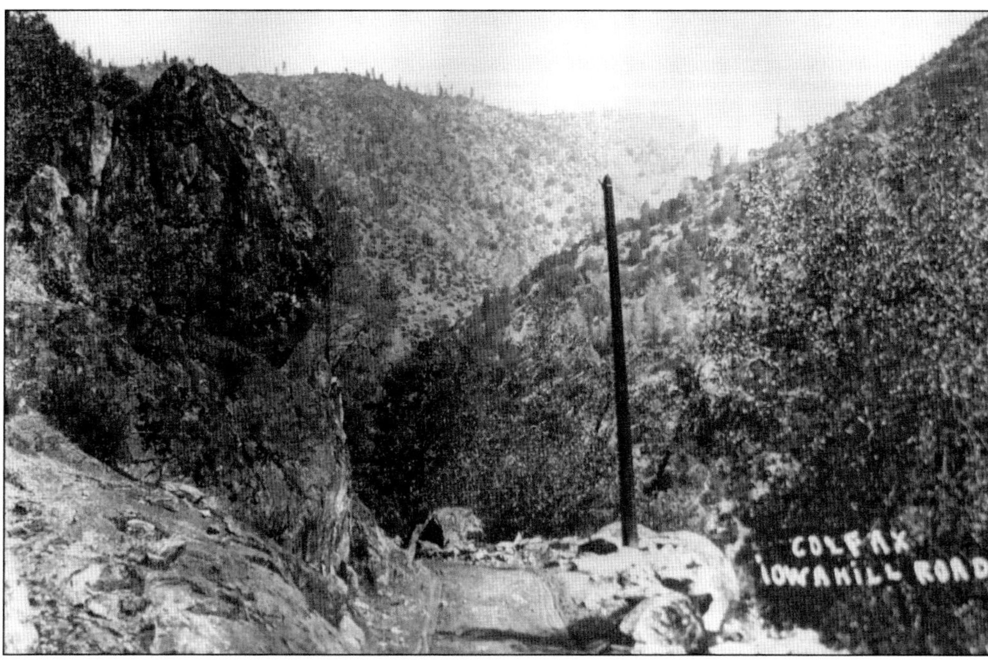

The notorious Colfax–to–Iowa Hill Road, about one wagon wide, drops off to the American River below. Boulders loom above and alongside, threatening the confidence of unfamiliar travelers who pray they do not meet anyone coming the opposite way. Pioneers used the Dinner Tree, halfway between Colfax and Iowa Hill, as their resting point along the stage route. Though paved now, the road still has a reputation for being treacherous. (CHM.)

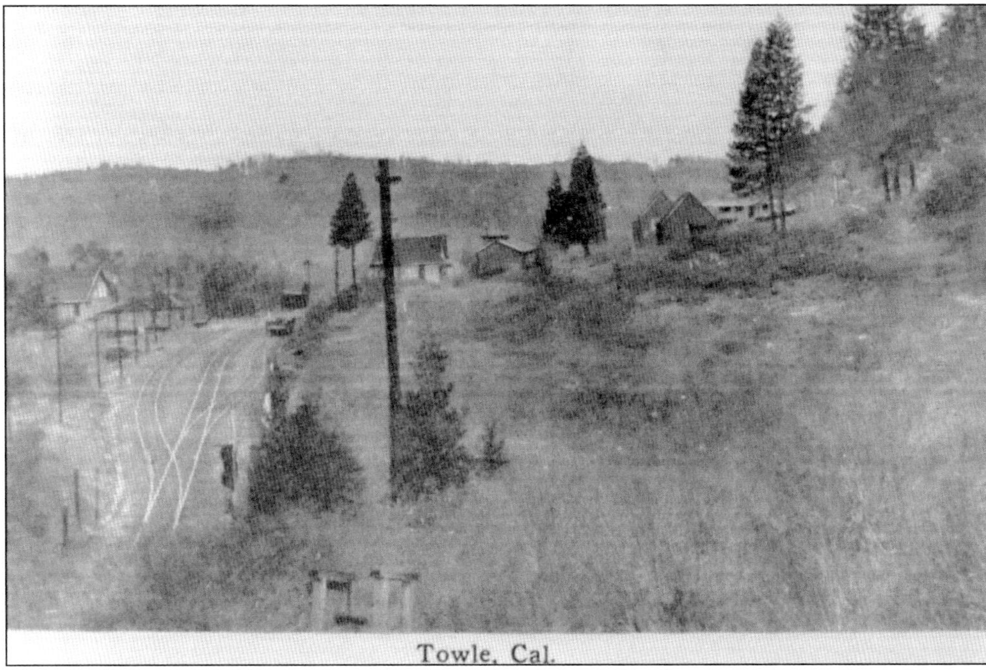

The Towle brothers began their timber business in the town of Dutch Flat. In 1859, they established their first sawmill at Lost Camp, where Blue Canyon is today. Later, they moved up the hill and managed other lumber mills and their own private narrow-gauge train from their new community of Towle on the transcontinental railroad line. (CHM.)

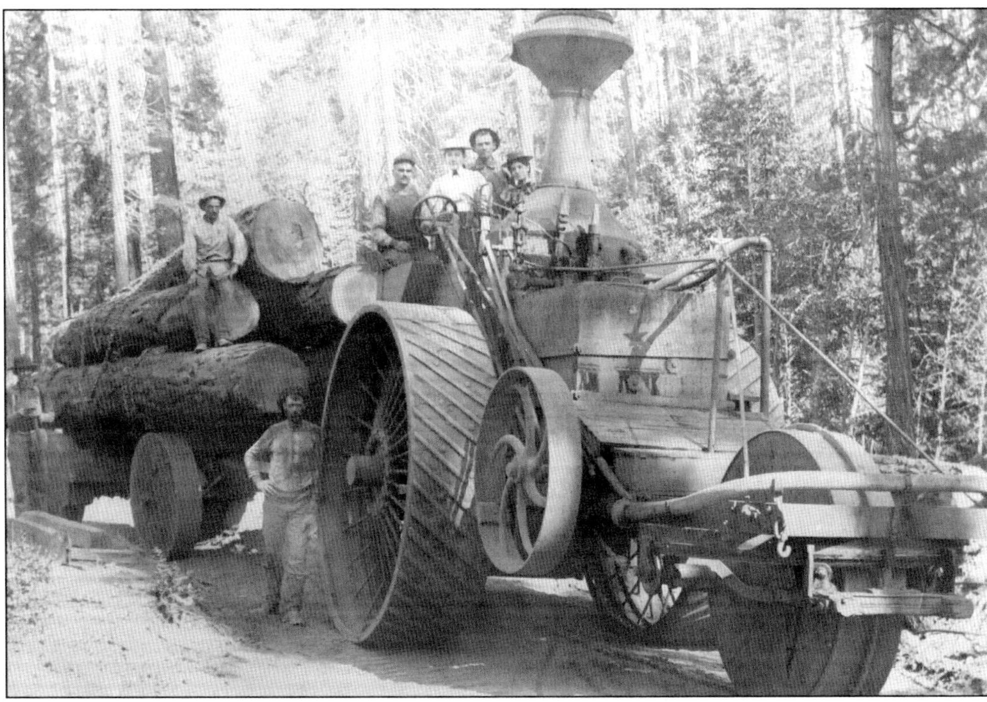

This large steam wagon hauls timber to the sawmill at Texas Hill. The Towle brothers and others provided lumber to the railroads and the building and mining industries. (CHM.)

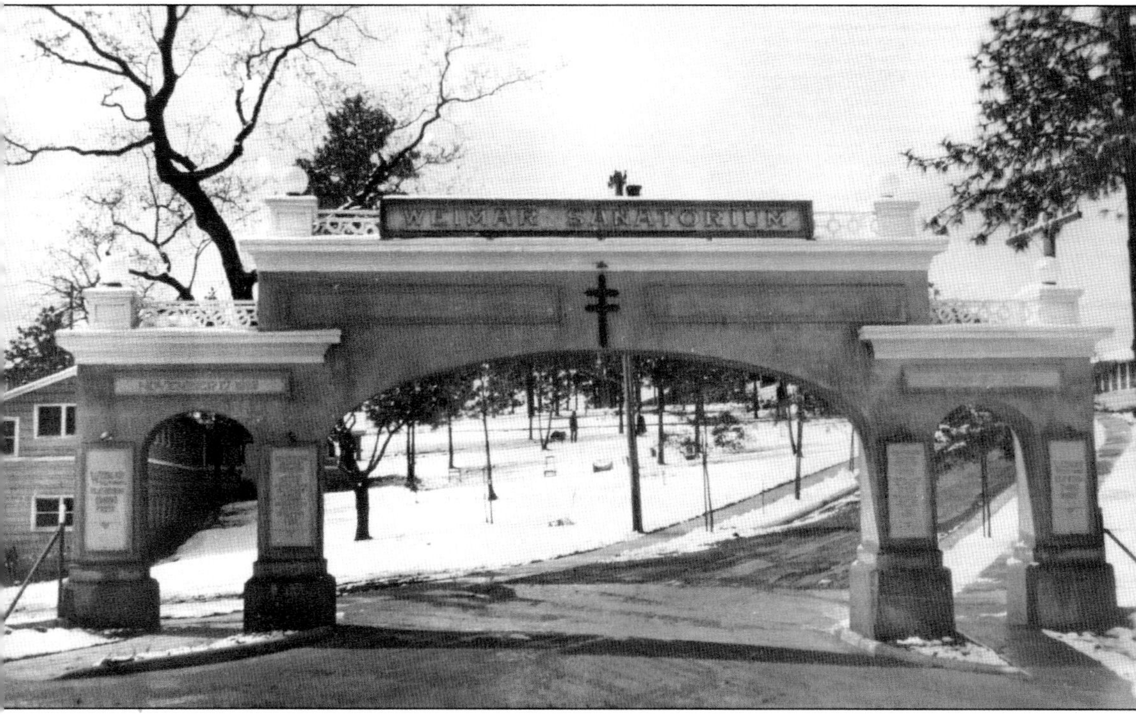

The town of Weimar became the site of Weimar Sanatorium, renowned for its work with tuberculosis patients. Dr. Robert Peers, who arrived in Colfax in 1899, became an advisor for the new facility, which was constructed just six miles down from Colfax on old Highway 40. Dr. Peers saw the climate of the Sierra foothills as ideal for treating patients. His process involved isolating patients from the populace and placing them in a quiet facility with constantly circulating fresh air. The Weimar Sanatorium opened on November 17, 1919. After the number of tuberculosis cases abated, the facility treated the mentally ill. Then in 1960, it was renamed the Weimar Medical Clinic. In 1966, it became a general hospital, but it closed for financial reasons in 1972. After having different owners, it became a temporary housing center for Vietnamese refugees and was eventually purchased by Seventh-Day Adventists. These three arches no longer stand. (AS.)